HORTICULTURE EDITION

Western Fertilizer Handbook

Horticulture enhances our enviroment.

HORTICULTURE EDITION

Western Fertilizer Handbook

Produced by

**SOIL IMPROVEMENT COMMITTEE
CALIFORNIA FERTILIZER ASSOCIATION**

INTERSTATE PUBLISHERS, INC.

DANVILLE, ILLINOIS

WESTERN FERTILIZER HANDBOOK
Horticulture Edition

Copyright © 1990 by

California Fertilizer Association
1700 I Street, Suite 130
Sacramento, CA 95814

Library of Congress Catalog Card No. 89-85732

ISBN 0-8134-2858-0 (paperbound)

2 3
4 5 6
7 8 9

Order from

Interstate Publishers, Inc.

510 North Vermilion Street
P.O. Box 50
Danville, IL 61834-0050

Phone: (800) 843-4774
Fax: (217) 446-9706

Acknowledgements

This is the first edition of the *Western Fertilizer Handbook* prepared specifically for the horticulture sector of the fertilizer industry. The book presents fertilization, nutrient management and related topics based on the fundamentals of biological and physical sciences, as does each edition of the original *Handbook*. In this case the information is set in a horticultural context in the hope that it can be more easily understood by the student and more rapidly applied by the professional as well as the amateur horticulturist.

Much of the information in this horticulture version was taken directly from the *Western Fertilizer Handbook,* 7th ed. The Editorial Committee, therefore, gratefully acknowledges the original contributions of the many excellent scientists to the original and, therefore, to the horticulture edition of the *Western Fertilizer Handbook.*

The Editorial Committee converted the text and added original information. Many other members of the California Fertilizer Association and colleagues of the association contributed ideas and photos. They are also gratefully acknowledged.

EDITORIAL COMMITTEE
Albert E. Ludwick, Chairman
Keith B. Campbell
Richard D. Johnson
Lynette J. McClain
Robert M. Millaway
Steven L. Purcell
Irvine L. Phillips, Jr.
Dale W. Rush
Joe A. Waters

Table of Contents

Introduction

Western horticulture may be described in many ways — dynamic, aggressive, modern, changing, diverse. . . . Certainly its trademark is the production of a wide range of high-quality plant products. The unique climate of the West, moderated by the Pacific Ocean and influenced by the mountain ranges, is really a broad spectrum of micro-climates, allowing, and indeed encouraging, the very diversity in the horticulture industry that exists today.

The fertilizer industry contributes significantly to this production of high-quality horticultural products and has done so for many years. It is proud of its role and looks forward to continuing success.

The *Western Fertilizer Handbook — Horticulture Edition* is written specifically for western horticulture and the individual seeking practical problem-solving information. Its language is concise and straightforward. Information is included on properties of soil, water and plants. Fertilizer products, their properties and their management are presented in this book, as is related information on soil amendments and water quality.

HOW PLANTS GROW

A knowledge of how plants grow and how they depend on soil to sustain them is key to becoming a successful horticulturist. This information will form the basis for many management decisions related to fertilization and irrigation. Proper decisions will lead to excellent end products, whether they be flowers, turf, vegetables, ornamental trees and shrubs, tree and vine crops or others.

It can be said that the climate in the West is far superior to its soils. Soils ideal for optimum plant growth are rare. In fact, when such a soil is encountered it is usually the result of careful management and the addition of appropriate amendments over a period of years. It is, therefore, necessary to have an understanding of the basic chemical and physical components of soil and how these components can be modified to produce the desired result. Problems frequently dealt with include alkalinity, salinity, acidity, nutrient deficiency, and shallow rooting caused by hard pans or other internal soil barriers. Some problems encountered are fairly obvious and easy to correct; others will require professional help.

The other ingredient to consider is water. "How often should I water?" is perhaps the question most frequently asked by the novice. This is a very difficult question to answer. The answer will depend on many variables, including needs of the particular plant, rooting habit, age, temperature, wind, day length, method of irrigation and even the quality of the water itself. Ignoring these factors and watering by the calendar and the clock will seldom give satisfactory results. Invariably, plants will become subject by stress from drought or drowning. Their growth will slow and their appearance will deteriorate. Such plants are also more subject to attack by insects and diseases.

FERTILIZATION

There is an old saying that sums up the role of fertilizer very well: "A fertile soil is not always a productive soil, but a productive soil is always a fertile soil." Fertilizer is food for plants. When this food is not provided or when the individual nutrients are out of balance, the plant starves. Nutrient starvation results in slow growth, chlorosis (yellowing) and finally death of the plant. Proper fertilization which supplements what the soil may supply is essential for optimum growth and quality.

Each of the 16 essential nutrients is required in different amounts by plants. This is complicated further by the fact that when applied to the soil (Note: carbon, hydrogen and oxygen are supplied from air and water), nutrients react differently. Some are soluble in water, others precipitate, some react with clay particles or organic matter, etc. Again, a basic understanding of soil (both chemical and physical properties) will help sort out the differences and will help to determine how to manage each nutrient.

A rose by any other name is still a rose (another old saying). Just as true, a nutrient is a nutrient regardless of its source. Plants do not distinguish between nutrient sources. Nitrogen from manure is absorbed by plant roots as ammonium or nitrate just as it is from commercial ammonium nitrate that has been applied. Environmental quality is important and certainly worth protecting. We can do this best by knowing the facts and making well-thought-out decisions based on those facts, not on fantasy, fear or ignorance.

The purpose of the *Western Fertilizer Handbook — Horticulture Edition*, as previously stated, is to present the facts in a concise and straightforward manner. The "facts" are derived primarily from many years of research and experience by your university and government researchers, extension specialists and industry professionals. Study the information and use it to the best of your ability.

CALIFORNIA FERTILIZER ASSOCIATION

The California Fertilizer Association was established in 1923 for the purpose of promoting progress within the fertilizer industry in the interest of an efficient and profitable agriculture/horticulture community. This continues to be its purpose today.

Activities of the association include developing and disseminating new information to its members and others; encouraging environmentally safe and efficient use of fertilizer; supporting production-oriented research programs to identify maximum economic yields for farmers and horticulturists; promoting the teaching of agronomic/horticultural topics in our schools, colleges and universities; and maintaining open lines of communication among industry, university and other state as well as federal agencies.

Many of the above activities are carried out by the Soil Improvement Committee, which was established by the membership in the mid-1920s. The Soil Improvement Committee produced the first edition of the *Western Fertilizer Handbook* in 1953, and with later editions, total distribution now exceeds 150,000 copies. This committee also produced this book—the first horticulture edition.

There are other professional organizations that sponsor programs similar to those of the California Fertilizer Association. A partial list is included at the end of Chapter 13.

CHAPTER 1

Soil — A Medium for Plant Growth

Food comes from the earth. The land with its waters gives us nourishment. The earth rewards richly the knowing and diligent but punishes inexorably the ignorant and slothful. This partnership of land and grower is the rock foundation of our complex social structure.

W. C. Lowdermilk

WHAT IS SOIL?

As a medium for plant growth, soil can be described as a complex natural material derived from disintegrated and decomposed rocks and organic materials, which provides nutrients, moisture and anchorage for land plants.

The four principal components of soil are mineral materials, organic matter, water and air. These are combined in widely varying amounts in different kinds of soil, and at different moisture levels. A representative western soil, at an ideal moisture content for plant growth, is nearly equally divided between solid materials and pore space, on a volume basis. The pore space contains nearly equal amounts of water and air. Figure 1-1 shows a schematic representation of such a soil.

HOW SOILS ARE FORMED

The development of soils from original rock materials is a long-term process involving both physical and chemical weathering, along with biological activity. The widely variable characteristics of soils are due to differential influences of the soil formation factors:

1. Parent material — material from which soils were formed.
2. Climate — temperature and moisture.

1

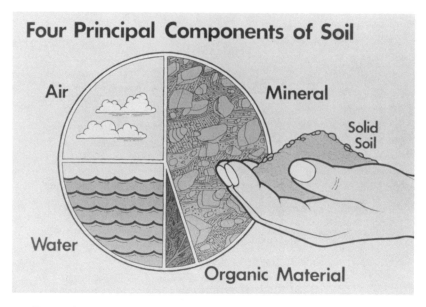

Fig. 1-1. Volumetric content of four principal soil components for a representative western soil at an ideal moisture content for plant growth.

3. Living organisms — microscopic and macroscopic plants and animals.
4. Topography — shape and position of land surfaces.
5. Time — period during which parent materials have been subjected to soil formation.

The initial action on the original rock is largely mechanical — cracking and chipping due to temperature changes. As the rock is broken into smaller particles, the total surface area exposed to the atmosphere increases. Chemical action of water, oxygen, carbon dioxide and various acids further reduces the size of rock fragments and changes the chemical composition of many of the resultant particulate materials. Finally, the action of microorganisms and higher plant and animal life contributes organic matter to the weathered rock material, and a true soil begins to form.

Since all of these soil-forming agents are in operation constantly, the process of soil formation is a continual one. Evidence indicates that the soils we depend on today to produce our crops required hundreds or thousands of years to develop. In this regard, we might consider soil as a nonrenewable resource, measured in terms of our life span. Thus it is very important that we protect our soils from destructive erosive forces and nutrient depletion, which can rapidly destroy the product of hundreds of years of nature's work.

SOIL PROFILE

A vertical section through a soil typically presents a layered pattern. This section is called a *profile,* and the individual layers are referred to as *horizons.* A representative soil has three general horizons, which may or may not be clearly discernible (Figure 1-2).

The uppermost horizon includes the *surface soil,* or *topsoil,* and is designated the *A* horizon. The next successive horizon, underlying the surface soil, is called the *subsoil,* or *B* horizon. The combined A and B horizons are referred to as the *solum.* Underlying the B horizon is the *parent material,* or *C* horizon. These three horizons, together with the unweathered, nonconsolidated rock fragments lying on top of the bedrock, are called the *regolith.*

Soil profiles vary greatly in depth or thickness, from a fraction of an inch to many feet. Normally, however, a soil profile will extend to a depth of about 3 to 6 feet. Other soil characteristics, such as color, texture, struc-

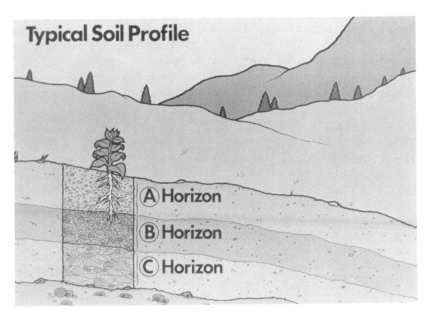

Fig. 1-2. This typical soil profile illustrates the three general soil horizons.

ture and chemical composition also exhibit wide variations among the many soil types.

The surface soil (A horizon) is the layer which is most subject to climatic and biological influences. Most of the organic matter accumulates in this layer, which usually gives it a darker color than the underlying horizons. Commonly this layer is characterized by a loss of soluble and colloidal materials, which are moved into the lower horizons by infiltrating water, a process called *eluviation*.

The subsoil (B horizon) is a layer which commonly accumulates many of the materials leached and transported from the surface soil. This accumulation is termed *illuviation*. The deposition of materials such as clay particles (colloidal material carried by infiltrating water from the surface layer, as well as synthesized clay particles formed from the recombining of soluble silicates and hydroxides), iron, aluminum, calcium carbonate, calcium sulfate and other salts creates a layer which normally has more compact structure than the surface soil. This often leads to restricted movement of moisture and air within this layer, which produces an important effect upon plant growth.

The parent material (C horizon) is the least affected by physical, chemical and biological agents. It is very similar in chemical composition to the original material from which the A and B horizons were formed. Parent material which has formed in its original position by weathering of bedrock is termed *sedentary* or *residual*, while that which has been moved to a new location by natural forces is called *transported*. This latter type is further characterized on the basis of the kind of natural force responsible for its transportation and deposition. When water is the transporting agent, the parent materials are referred to as *alluvial* (stream-deposited), *marine* (sea-deposited) or *lacustrine* (lake-deposited). Wind-deposited materials are called *aeolian*. Materials transported by glaciers are termed *glacial*. And, finally, those that are moved by gravity are called *colluvial*, a category that is relatively unimportant with respect to horticultural soils.

Because of the strong influence of climate on soil profile development, certain general characteristics of soils formed in areas of different climatic patterns can be described. For example, much of the western area has an arid climate, which results in the development of coarser-textured soils (more sand particles) than most of those developed in more humid climates. Also, the soil profiles in many western soils are less developed, since the amount of water percolating through the soils is generally much less than in more humid climates. Because of this, many western soils contain more calcium, potash, phosphate and other nutrient elements than do the more extensively developed eastern soils.

Thus, the soil profile is an important consideration in terms of plant growth. The depth of the soil, its texture and structure and its chemical nature determine to a large extent the value of the soil as a medium for plant growth.

SOIL TEXTURE

Soils are composed of particles with an infinite variety of sizes and shapes. On the basis of their size, individual mineral particles are divided into three categories — sand, silt and clay. Such a division is very meaningful, not only in terms of a classification system but also in relation to plant growth. Many of the important chemical and physical reactions are associated with the surface of the particles. Surface area is enlarged greatly as particle size diminishes, which means that the smallest particles (clay) are the most important with respect to these reactions.

Two systems of classification of the various particle sizes (soil separates) are used. One is the U.S. Department of Agriculture system, and the other is the International system. The size ranges of the various soil separates for the two systems are shown in Table 1-1.

Soil texture is determined by the relative proportions of sand, silt and clay found in the soil. Twelve basic soil textural classes are recognized. A classification chart based on the actual percentages of sand, silt and clay appears in Figure 1-3.

Table 1-1

Size Limits of Soil Separates[1]

U.S. Department of Agriculture		International	
Name of Separate	Diameter (Range)	Fraction	Diameter (Range)
	(mm)		*(mm)*
Very coarse sand	2.0 – 1.0		
Coarse sand	1.0 – 0.5	I	2.0 – 0.2
Medium sand	0.5 – 0.25		
Fine sand	0.25 – 0.10	II	0.20 – 0.02
Very fine sand	0.10 – 0.05		
Silt	0.05 – 0.002	III	0.02 – 0.002
Clay	Below 0.002	IV	Below 0.002

[1]Source: USDA *Soil Survey Manual*, Agricultural Handbook No. 18.

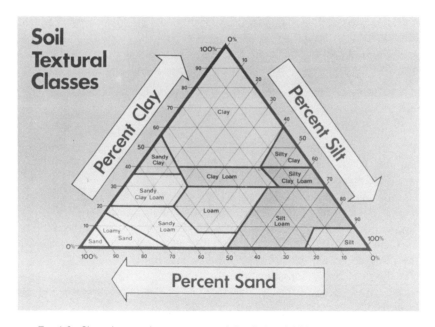

Fig. 1-3. Chart showing the percentages of clay (below 0.002 mm), silt (0.002 to 0.05 mm) and sand (0.05 to 2.0 mm) in the basic soil textural classes.

A textural class description of soils reveals a lot about soil-plant inter-actions, since the physical properties of soils are determined largely by the texture. In mineral soils, the exchange capacity (ability to hold plant nutri-ent elements) is related closely to the amount and kind of clay in the soil. The water-holding capacity is determined in large measure by the particle size distribution. Fine-textured soils (high percentage of silt and clay) hold more water than coarse-textured soils (sandy). Finer-textured soils often are more compact, have slower movement of water and air and can be more difficult to work.

From the standpoint of plant growth, medium-textured soils, such as loams, sandy loams and silt loams, are probably the most ideal. Neverthe-less, the relationships between soil textural class and soil productivity can-not be applied generally to all soils, since texture is only one of the many factors that influence plant growth.

SOIL STRUCTURE

Except for sand, soil particles normally do not exist singularly in the soil but rather are arranged into aggregates, or groups of particles. The way in which particles are grouped together is termed *soil structure*.

There are four primary types of structure, based upon shape and arrangement of the aggregates (see Figure 1-4). Where the particles are arranged around a horizontal plane, the structure is called *plate-like* or *platy.* This type of structure can occur in any part of the profile. Puddling or ponding of soils often gives this type of structure on the soil surface. When particles are arranged around a vertical line, bounded by relatively flat vertical surfaces, the structure is referred to as *prism-like* (prismatic or columnar). Prism-like structure is usually found in subsoils and is common in arid and semiarid regions. The third type of structure is referred to as *block-like* (angular blocky or subangular blocky) and is characterized by approximately equal lengths in all three dimensions. This arrangement of aggregates is also most common in subsoils, particularly those in humid regions. The fourth structural arrangement is called *spheroidal* (granular or crumb) and includes all rounded aggregates. Granular and crumb structures are characteristic of many surface soils, particularly where the organic matter content is high. Soil management practices can have an important influence on this type of structure.

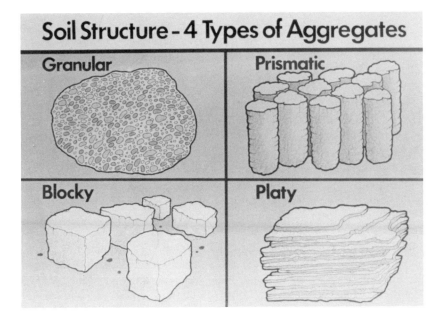

Fig. 1-4. Generalized illustration of the four types of soil aggregates.

Soil aggregates are formed both by physical forces and by binding agents — principally products of decomposition of organic matter. The latter types are more stable and resist to a greater degree the destructive forces of water and cultivation. Aggregates formed by physical forces such as drying, freezing and thawing, and tillage operations are relatively unstable and are subject to quicker decomposition.

Soil structure has an important influence on plant growth, primarily as it affects moisture relationships, aeration, heat transfer and mechanical impedance of root growth. For example, the importance of good seedbed preparation is related to moisture and heat transfer — both of which are important in seed germination. A fine granular structure is ideal in this respect.

The movement of moisture and air through the soil is dependent upon the porosity, which is influenced markedly by soil structure. Granular structure provides adequate porosity for good infiltration of water and air exchange between the soil and the atmosphere. This creates an ideal physical medium for plant growth. However, where surface crusting occurs, or subsurface claypans or hardpans exist, plant growth is hindered because of restricted porosity. Good management practices can improve soil structure and thereby create a better condition for plant growth.

SOIL REACTION (pH)

The terms *acid, neutral* and *alkaline* refer to the relative concentrations of hydrogen ions (H^+) and hydroxyl ions (OH^-) in the soil solution. These concentrations are measured in terms of a pH value, which gives a measure of the active acidity in the soil solution, rather than the total or potential acidity of the soil. An acid soil has a higher concentration of hydrogen ions than hydroxyl ions, while an alkaline soil has the opposite. A neutral pH means that the two kinds of ions are present in equal amounts and counteract the effect of one another.

In order to distinguish between the relative degrees of acidity or alkalinity of a soil, a pH scale from 0 to 14 is used. At the middle of the scale (pH 7.0), soil is neutral in reaction, while below 7.0 the reaction is acidic, and above 7.0 it is alkaline (basic). The lower the pH value, the more acid the soil, and conversely, the higher the pH, the more alkaline. Since pH is a logarithmic function, each pH unit represents a tenfold increase or decrease in relative acidity or alkalinity. For example, a soil with a pH of 6.0 is 10 times as acid as one with a pH of 7.0. Also, a soil with a pH of 8.0 is

10 times as alkaline as one with a pH of 7.0 and 100 times as alkaline as one with a pH of 6.0.

The soil reaction is important to plant growth for several reasons: (1) its effect on nutrient availability, (2) its effect on the solubility of toxic substances (such as aluminum), (3) its effect on soil microorganisms and (4) the direct effect of pH on root cells (which affects the uptake of nutrients and water).

The availability of each plant nutrient varies at different pH levels. A pH between 6.5 and 7.5 gives maximum availability of the primary nutrients (N-P-K) and a relatively high availability of the other nutrients. For most plants, a soil pH of 6.0 to 7.0 is the most satisfactory range.

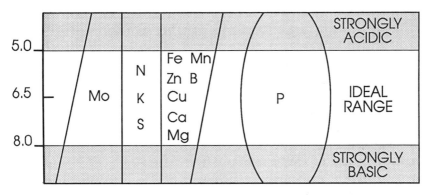

Fig. 1-5. Relative mineral nutrient availability as affected by soil pH.

Soils become more acid as a result of leaching the cations calcium, magnesium and potassium from the topsoil into the subsoil, and through removal of the cations by growing crops. As the cations are removed from the soil particles, they are replaced with acid-forming hydrogen and aluminum ions.

Calcium carbonate in the soil acts as a buffer against acid formation, which means that it tends to restrict the development of acid soils. This occurs due to the increased solubility of calcium carbonate as the acidity increases, which increases the exchangeable calcium in the soil and removes hydrogen ions, which combine with oxygen from the carbonate to form water. Carbon dioxide is released in the process. For this reason, calcium carbonate, or agricultural limestone, is used as an amendment on acid soils (see Tables 7-2 and 7-3).

In many regions soils are naturally alkaline (high pH), which may affect nutrient availability. In such cases elemental sulfur is used to lower pH. Soil microorganisms (*Thiobacillus* spp.) convert the sulfur to sulfuric acid. The sulfuric acid produced in the reaction is ionized and releases hydrogen ions (H^+), which lower pH.

Most ammonical and urea forms of nitrogen fertilizers contribute to soil acidity, since their reactions in the soil increase the concentration of hydrogen ions in the soil solution. The use of these nitrogen fertilizers increases the acidulation process in soils. This can improve nutrient uptake by plants in high-pH soils or decrease nutrient uptake in low-pH soils. This adds to the importance of soil testing to determine if corrective measures might become necessary.

CATION EXCHANGE CAPACITY

Due to their chemical structure, clay particles and decomposed soil organic matter (humus) are electrically active and carry a net negative charge. This means that electrically charged ions that carry a positive charge (cations) can be attracted to and held by these soil materials. Cations in the soil solution or adsorbed on the surface of plant roots can exchange positions with those adsorbed on the surface of clays or humus materials. Figure 1-6 illustrates this exchange. The cation exchange capacity of a soil is a measure of the quantity of such cations that can be adsorbed or held by a soil.

Since soils contain varying amounts and different kinds of clay and humus, the total exchange capacity is widely variable among soil types. Most western soils contain predominantly the montmorillonitic and hydrous mica type clays, which have an exchange capacity approximately 5 to 10 times as great as that of the kaolinitic clays typical of soils in the southeastern United States. Because of the high exchange capacity of soil humus materials, soils with a high percentage of organic matter typically have higher exchange capacities than those of low organic matter content with similar amounts and types of clay.

The cations of greatest significance with respect to plant growth are calcium (Ca^{++}), magnesium (Mg^{++}), potassium (K^+), ammonium (NH_4^+), sodium (Na^+) and hydrogen (H^+). The first four are plant nutrients and are, therefore, involved directly with plant growth. The last two have a pronounced effect upon nutrient and moisture availability.

The relative amount of each of the cations adsorbed on clay particles is closely associated with important soil properties. Highly acid soils have a

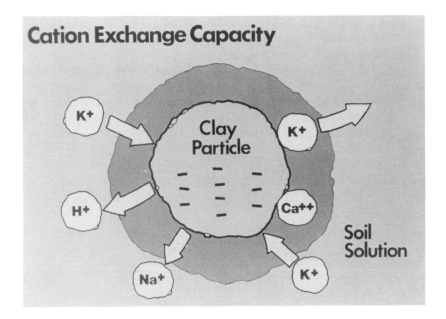

Fig. 1-6. Schematic illustration of the exchange of cations between the negatively charged clay particles and the soil solution.

high percentage of adsorbed hydrogen, while soils with a favorable pH (6.0 to 8.0) have predominantly calcium ions. Soils that are high in sodium ions are dispersed and resist the infiltration of water, while those with a high percentage of calcium ions are flocculated and favor higher infiltration rates. Mineral soils with a high exchange capacity are typically more fertile than those with lower exchange capacities, since they resist the loss of plant nutrients through the leaching process.

SOIL MICROORGANISMS

Besides their role in soil-forming processes, soil organisms make an important contribution to plant growth through their effect on the fertility level of the soil. Particularly important in this respect are the microscopic plants (microflora) which function in decomposing organic residues and releasing available nutrients for growing plants.

Some important kinds of microorganisms are bacteria, fungi, actinomycetes and algae. All of these are present in the soil in very large numbers, when conditions are favorable. A gram of soil (about 1 cubic centimeter) may contain as many as 4 billion bacteria, 1 million fungi, 20 million actinomycetes and 300,000 algae. These microorganisms are important in the decomposition of organic materials, the subsequent release of nutrient elements and the fixation of nitrogen from the atmosphere.

Soil bacteria are of special interest because of their many varied activities. In addition to the group of bacteria which function in decomposing organic materials (heterotrophic bacteria), there is a smaller group (autotrophic bacteria) which obtain their energy from the oxidation of mineral materials such as ammonium, sulfur and iron. This latter group is responsible for the nitrification process (oxidation of ammonium to nitrate nitrogen) in the soil, a process which is vitally important in providing nitrogen for the growth of plants.

Nitrogen-fixing bacteria also play an important role in the growth of higher plants, since they are capable of converting atmospheric nitrogen into useful forms in the soil. Nodule bacteria (rhizobia) live in conjunction with roots of leguminous plants, deriving their energy from the carbohydrates of the host plants, and fix nitrogen from the soil atmosphere. Free-living bacteria (azotobacter and clostridium) also fix atmospheric nitrogen, although to a lesser extent than the rhizobia bacteria, under most conditions.

Because of the important contributions made by bacteria to the fertility level of the soil, it has been stated that if their functions were to fail, life for higher plants and animals would cease.

SOIL MANAGEMENT

"To use to the best advantage" is the definition of the word *manage* which best applies to soil management. This implies using the best available knowledge, techniques, materials and equipment for growing plants. Through wise management, the grower can produce crops in the abundance required to feed the growing population, grow ornamentals and turf that add to the aesthetic value of our surroundings, and at the same time improve soils and the environment, thus providing a priceless legacy for future generations.

Tillage and soil amendments are important management practices used in growing plants. They serve many purposes — including seedbed preparation, weed control, incorporation of plant residues and fertilizer materials, breaking soil crusts and hardpans to improve water penetration and aeration, and improving the soil for irrigation and erosion control. Because of the potential damage to soil structure, the soil should not be over-worked. Tillage and soil amendments are determined by the type of plants, type of soil and local growing conditions. No one set of guidelines or standards is appropriate for all situations.

Soil conservation is another important management practice which deserves close attention. It is estimated that annually in the U.S., 4 billion tons of sediment are lost from the land in runoff waters. That is equivalent to the total loss of topsoil (6 inches deep) from 4 million acres. Wind erosion is also a problem in certain areas, particularly in arid regions. Manage-

Fig. 1-7. Careful soil and plant management is important . . . for today and tomorrow. (Photo is of commercial turf harvest.)

ment practices such as contouring, strip planting, windbreaks, cover-cropping, reduced tillage, terracing and plant residue management help to eliminate or minimize the loss of soil from water and wind erosion. These practices are valuable in horticultural situations as they have been in other types of agriculture. In addition to these practices, a sound fertilizer program promotes optimal growth of plants, which contributes to soil erosion control by protecting the soil against the impact of falling rain and by holding the soil in place with extensive plant root systems.

Proper utilization of plant residues can be a key management practice. Plant residues returned to the soil improve soil productivity through the addition of organic matter and plant nutrients. The organic matter also contributes to an improved physical condition of the soil, which increases water infiltration and storage and aids aeration. This is vital to plant growth, and it improves soil tilth. In deciding how to best utilize plant residues, the immediate benefits of burning or removal should be weighed against the longer-term benefits of soil improvement brought about by incorporation of residues into the soil.

Special consideration should be given to the environmental aspects of soil management. Comments regarding soil erosion in terms of soil losses have already been made. The environmental implications of erosion are extremely important, since sediment is by far the greatest contributor to water pollution. Management practices which minimize soil erosion losses, therefore, contribute to cleaner water. The judicious use of fertilizers, which includes using the most suitable analyses and rates of plant nutrients, as well as the proper timing of application and placement in the soil, is also important. Only when improperly used are fertilizers a potential pollution hazard. When used judiciously, they can make a significant contribution to a cleaner, more enjoyable environment. These effects are noted in three areas — cleaner air, cleaner water and improved wildlife habitat.

SUPPLEMENTARY READING

1. *Fundamentals of Soil Science,* Seventh Edition. H. D. Foth. John Wiley & Sons, Inc. 1984.
2. *Lawns — Basic Factors, Construction and Maintenance of Fine Turf Areas,* Third Edition. J. Vingris. Thomson Publications. 1982.
3. *The Nature and Properties of Soils,* Eighth Edition. N. C. Brady. The Macmillan Company. 1974.
4. *Soil Conservation,* Second Edition. N. Hudson. Cornell University Press. 1981.
5. *Soil Genesis and Classification,* Second Edition. S. W. Buol, F. D. Hole and R. J. McCracken. Iowa State University Press. 1980.
6. *Turfgrass: Science and Culture.* J. B. Beard. Prentice-Hall, Inc. 1973.
7. *Turfgrass Management,* Revised Edition. A. J. Turgeon. Reston Publishing Co. 1985.
8. *Turf Management for Golf Courses.* J. B. Beard. Burgess Publishing Co. 1982.

CHAPTER 2

Water and Plant Growth

All water used for irrigation contains some dissolved salts. The suitability of water for irrigation generally depends on the kinds and amounts of salts present. All salts in irrigation waters have an effect on plant-soil-water relations, on the properties of soils and indirectly on the production of plants.

A user of irrigation water should know the effects that water quality and irrigation practices have on:

1. The salt content (salinity) of the soil.
2. The sodium status of the soil.
3. The rate of water penetration into the soil.
4. The presence of elements which may be toxic to the plants.

It is difficult to isolate these factors one from another because some of them are interrelated.

IRRIGATION

Irrigation water is applied to soil to replenish the water removed from the soil by evaporation, by growing plants and to a lesser extent by drainage below the root zone. It is applied in a number of ways. The method of application depends primarily upon the plants to be grown, the depth and texture of the soil, the topography of the land and the cost of water. The amount of water used and how often it is applied are determined by the needs of the various plants and the need to provide deep leaching occasionally to prevent accumulation of salts within the root zone. Therefore, successful irrigation requires careful management of both plants and water.

SOIL MOISTURE BEHAVIOR

In a well-drained soil, water is held largely as a film around each soil

Fig. 2-1. Irrigation water is applied to replace water removed from the soil.

particle. The thinner these films are, the more tightly the water is held and the greater the suction needed to remove the water.

Immediately following an irrigation, the films of water are thick and, therefore, are not tightly held on the soil. This condition is called the *saturation percentage.*

In about two or three days, with free drainage, about one-half of this weakly held water moves deeper into the soil, and free drainage practically ceases. The moisture content at this point is called the *field moisture capacity.* The films of water are now thinner and are held more tightly.

Below the field moisture capacity, gravity is no longer a significant force in moving water in the soil. Most of the water is then removed by the roots of growing plants. Plants will remove about one-half of the water held at the field moisture capacity. At that point the soil holds moisture so tightly that plants cannot extract it, thus causing them to wilt. That point is called *permanent wilting percentage.*

The moisture content of a soil saturated with water, its *saturation percentage,* is about twice the field moisture capacity and about four times the permanent wilting percentage. This relationship between the saturation percentage, field moisture capacity and permanent wilting percentage is accurate for practical purposes for all soils from clay loams to sandy loams.

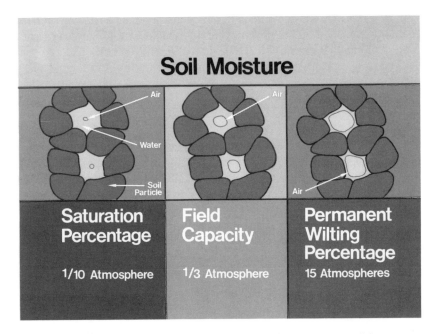

Fig. 2-2. Illustration of the cardinal soil moisture values — saturation percentage, field capacity and permanent wilting percentage.

Table 2-1

Approximate Amounts of Water
Held by Different Soils

Soil Texture	Inches of Water Held per Foot of Soil	Max. Rate of Irrigation — Inches per Hour, Bare Soil
Sand	0.5 – 0.7	0.75
Fine sand	0.7 – 0.9	0.60
Loamy sand	0.7 – 1.1	0.50
Loamy fine sand	0.8 – 1.2	0.45
Sandy loam	0.8 – 1.4	0.40
Loam	1.0 – 1.8	0.35
Silt loam	1.2 – 1.8	0.30
Clay loam	1.3 – 2.1	0.25
Silty clay	1.4 – 2.5	0.20
Clay	1.4 – 2.4	0.15

Plants consume, on the average, from 0.1 to 0.3 in. of rainfall or irrigation per day.

A very moist soil has a thick film of water and hence has low suction. A drier soil has a thin film of water and has a high suction. For this reason, water will move from a wet soil to a drier soil, but the rate of such movement is slow. Table 2-1 shows the water-holding capacity of different soils.

WHEN TO IRRIGATE

Plant appearance is often used as a guide in determining when to irrigate. Symptoms such as slow growth, "bluish" color of leaves and temporary afternoon wilting are signs of moisture stress. Usually plants should be irrigated before these signs are conspicuous.

A soil tube or an auger can show depth of wetting and depletion of moisture. Experienced growers can determine fairly accurately the available moisture by feeling soil samples taken with a tube or an auger. A more accurate moisture determination can be made by placing weighed samples of moist soils in an oven to determine the percentage of moisture loss.

Moisture-measuring instruments such as tensiometers and electrical resistance blocks indirectly measure soil moisture and are used extensively

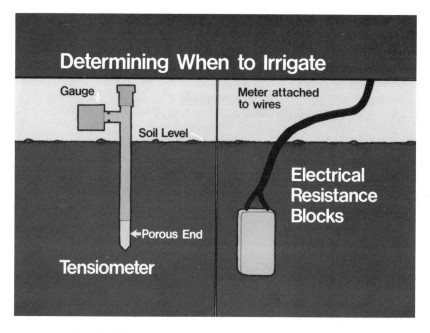

Fig. 2-3. Two moisture-measuring and -monitoring instruments.

in research work. Properly placed, they can be useful guides to determine the moisture levels in the soil and to schedule irrigations. Figure 2-3 illustrates two moisture-measuring and -monitoring instruments.

A *tensiometer* consists of a porous ceramic cup imbedded in the major root zone of the soil. A rigid tube is connected to the cup and to a vacuum gauge just above the soil surface. The whole system is filled with water and sealed tightly. As the roots remove water from the soil, the soil draws water from the porous cup, creating a suction which is measured by the gauge. The drier the soil, the greater the suction. Tensiometers are effective for suctions from 0 to 0.8 atmosphere (80 centibars on most gauges).

Electrical resistance blocks consist of one or more pairs of electrodes imbedded in a porous material, usually gypsum. The blocks are buried in the root zone with wires leading to the soil surface. As the soil moisture changes, the moisture in the gypsum block changes with it. When the soil is wet, the electrical resistance is low. As the soil and block become dry, the resistance increases.

This resistance is read by a specially designed meter attached to the wires. The meter is small and easily portable and can be used to monitor many stations. Gypsum blocks are usually not effective at soil tensions less than 2 atmospheres. Soluble salts in the soil solution decrease the electrical resistance and indicate a higher moisture content than is actually present.

Although tensiometers and resistance blocks are useful, they are limited to moisture readings in predetermined locations and depths only. These readings need to be made at regular, frequent intervals.

The number of instruments needed, the depth of placement and the interpretation of the readings depend upon the soil variability and upon the crop and irrigation system. Measurements of soil moisture levels provide only indirect measurements of plant moisture stress. Recently, some sensors have been developed to measure directly plant moisture stress in the field. These sensors can be used with confidence on some plants to determine the optimum time of irrigation. Hopefully, future research will refine the techniques to make these measurements more widespread.

WATER ANALYSIS TERMINOLOGY

Various terms and units are used in reporting a chemical analysis of water. An understanding of the commonly used methods of reporting is needed to interpret the data properly.

Dissolved salts are dissociated into electrically charged particles called *ions*. The ions which are positively charged are called *cations,* while the negatively charged ions are called *anions*. The concentration of each of these types of ions in the water is the usual method of reporting.

The common cations reported by laboratories are calcium (Ca^{++}), magnesium (Mg^{++}), sodium (Na^+) and potassium (K^+). The anions usually reported are bicarbonate (HCO_3^-), carbonate ($CO_3^=$), chloride (Cl^-) and sulfate ($SO_4^=$).

Anions usually found in smaller amounts are borate ($BO_3^=$) and nitrate (NO_3^-), which is sometimes reported as nitrate-nitrogen ($NO_3^- - N$).

Three principal methods are used to express the concentration of constituents in water. They are grains per gallon, parts per million and milliequivalents per liter.

Grains per gallon uses the English system of units. Most laboratories no longer use this method of reporting for irrigation water. It is still used to report hardness of water. To change grains per gallon to parts per million, multiply grains per gallon by 17.2.

Parts per million (ppm) is defined as 1 part of a salt to 1 million parts of water — or 1 milligram of salt per kilogram of solution. Water analyses are usually reported on a volume basis. One liter (1.057 quarts) of water weighs 1 kilogram (1,000 grams or 2.2 pounds); therefore, 1 milligram (one-thousandth of a gram) per liter is 1 part per million. Parts per million is used to report boron, nitrates, nitrate-nitrogen and a few other materials found in relatively small amounts.

Milliequivalents per liter (me / L) is the most meaningful method of reporting the major constituents of water. This is a measure of the chemical equivalence of an ion.

Salts are combinations of cations (sodium, calcium, magnesium, etc.) and anions (chlorides, sulfates, bicarbonates, etc.) in definite weight ratios. These weight ratios are based upon the atomic weight of each constituent and upon the valence (electrical charge). An equivalent weight of an ion is its atomic weight divided by its valence. A milliequivalent is $1/1,000$ equivalent. (See Table 2-2 for the approximate equivalent weights of the common ions.)

Some laboratories report the common constituents in parts per million. To convert to milliequivalents per liter, divide the ppm by the equivalent weight of an ion. For example, a laboratory reports 54 ppm of sulfate. The equivalent weight of sulfate found in Table 2-2 is 48. Divide 54 by 48 and find the water has 1.12 me / L of sulfate. (See Table 2-3 for water analysis conversions.)

Table 2-2

Major Constituents in Irrigation Water

	Ion Name	Symbol	Equivalent Weight
			(g)
Cations	Calcium	Ca^{++}	20
	Magnesium	Mg^{++}	12
	Sodium	Na^+	23
	Potassium	K^+	39
Anions	Bicarbonate	HCO_3^-	61
	Carbonate	$CO_3^=$	30
	Chloride	Cl^-	35.5
	Sulfate	$SO_4^=$	48

Table 2-3

Easy Conversion of Water Analysis of Ions to Pounds Material per Acre-Inch of Water

Ions	Milliequivalents per Liter (ppm)[1] (A)	Pounds per Acre-Inch of Water	Conversion Factor[2] (B)	Equivalent to Material in Pounds per Acre-Inch (C)	
Ca	20.0	4.54	0.57	11.4	$CaCO_3$
Mg	12.2	2.77	0.78	9.5	$MgCO_3$
Mg	12.2	2.77	1.12	13.7	$MgSO_4$
Na	23.0	5.22	0.58	13.3	NaCl
K	39.1	8.88	0.28	10.9	K_2O
K	39.1	8.88	0.43	16.8	KCl
CO_3	30.0	6.81	—	—	—
HCO_3	61.0	13.85	—	—	—
Cl	35.5	8.06	—	—	—
SO_4	48.0	10.90	0.075	3.6	S
SO_4	48.0	10.90	0.41	19.7	gypsum

[1]To convert Me / L to lbs. per acre-inch of water, multiply by 0.227.
[2]A $\times$ B = C.

The *pH* expresses the acidity or alkalinity of water. A pH reading of less than 7.0 is acidic, 7.0 is neutral and above 7.0 is alkaline. Most well waters range from pH 7.0 to pH 8.5. Some stream waters may be as acidic as pH 6.5. The pH measurement is not important if the major ions are reported and is thus often omitted from laboratory reports.

Total salt content is usually reported as the electrical conductivity (EC). Chemically pure water does not conduct electricity, but water with salts dissolved in it does. The more salt in the water, the better conductor it becomes. The ability of a water sample to conduct electricity is used to determine its salt content. EC is usually reported as decisiemens per meter (dS / m).[1]

The total salt content or total dissolved solids (TDS) is usually reported as ppm. This can be determined by evaporating a known weight of water sample to dryness and weighing the salt remaining. More often it is approximated by measuring the EC in dS / m and multiplying by 640. This gives the total dissolved solids in ppm. If a water has an EC of 1.6, then 1.6 $\times$ 640 = 1,024 ppm total dissolved solids. The me / L of total salts can be closely estimated by multiplying the EC in dS / m by 10. For example, a water with an EC of 2.62 dS / m contains 26.2 me / L of total salts.

Electrical conductivity is commonly used to check the salt content of soils. The conductivity is measured on the saturation extract from the soil and is designated EC_e. This is used to monitor the changes in the salt content of the soil resulting from irrigation. It is also useful in evaluating the relative tolerance of plants to salt and the suitability of a soil for certain crops.

Percent sodium is sometimes used. It is the ratio of sodium to the total cations in milliequivalents. A high sodium percentage may indicate a poor-quality water, but in recent years more refinement in interpretation of water quality has made this measurement less useful.

SOIL PROPERTIES AND WATER QUALITY

Over a long period of time the quality of irrigation water and the irrigation practice reach an equilibrium with the soil.

---

[1]Formerly referred to as millimhos per centimeter (mmhos / cm) or millisiemens per centimeter (mS / cm).

The Cations

To a great extent, the cations determine the physical as well as chemical properties of soil. The cations of most concern are calcium, magnesium, sodium and potassium.

Calcium (Ca^{++}) is found in essentially all natural waters. A soil well supplied with exchangeable calcium is friable and easily tilled and usually permits water to penetrate readily. Normally such soils do not "puddle" when wet. For these reasons, calcium in the form of gypsum is often applied to "tight" soils to improve their physical properties. Calcium replaces the sodium on the soil colloids and allows the sodium to be leached below the root zone. Generally, an irrigation water containing predominantly calcium is most desirable.

Magnesium (Mg^{++}) is also usually found in measurable amounts in most natural waters. Magnesium behaves much like calcium in the soil. Often laboratories will not separate calcium and magnesium but will report simply Ca + Mg in me / L. For most purposes this is adequate.

Sodium (Na$^+$) salts are all very soluble and as a result generally are found in all natural waters.

A soil with a large amount of sodium associated with the clay fraction has poor physical properties for plant growth. When it is wet, it runs together, becomes sticky and is nearly impervious to water. When it dries, hard clods form, making it difficult to till. Continued use of water with a high proportion of sodium may bring about these severe changes in an otherwise good soil. Eventually a sodic soil may develop. The so-called "slick spots" are usually spots high in exchangeable sodium.

Potassium (K$^+$) is usually found in only small amounts in natural waters. It behaves much like sodium in the soil. It is not often reported separately in water analyses but is included with the sodium.

The Anions

The anions indirectly affect the physical properties of soil by altering the ratio of calcium and sodium attached to the clays. The important anions are bicarbonate, carbonate, chloride and sulfate.

Bicarbonate (HCO$_3^-$) is common in natural waters. It is not usually found in nature except in solution. Sodium and potassium bicarbonates can exist as solid salts. An example is baking soda (sodium bicarbonate). Calcium and magnesium bicarbonates exist only in solution. As the moisture in the soil is reduced, either by removal by plants or by evaporation, calcium bicarbonate decomposes, carbon dioxide (CO_2) goes off into the air and water (H_2O) is formed, leaving insoluble lime ($CaCO_3$) behind.

$$Ca(HCO_3)_2 \xrightarrow{\text{upon drying}} CaCO_3 + CO_2 + H_2O$$

A similar reaction takes place with magnesium bicarbonate. Bicarbonate ions in the soil solution will precipitate calcium as the soil approaches dryness. This removes calcium from the clay and leaves sodium in its place. In this way, a calcium-dominant soil can become a sodium-dominant (sodic) soil by the use of a high-bicarbonate irrigation water.

Carbonate ($CO_3^=$) is found in some waters. Since calcium and magnesium carbonates are relatively insoluble, the cations associated with high-carbonate waters are likely to be sodium and possibly a small amount of potassium. When the soil dries, the carbonate ion will react with calcium and magnesium from the clay similar to bicarbonate, and an alkali (sodic) soil will develop.

Chloride (Cl^-) is found in all natural waters. In high concentrations it is toxic to some plants. All common chlorides are soluble and contribute to the total salt content (salinity) of soils. The chloride content must be determined to properly evaluate irrigation waters.

Sulfate ($SO_4^=$) is abundant in nature. Sodium, magnesium and potassium sulfates are readily soluble. Calcium sulfate (gypsum) has a limited solubility. Sulfate has no characteristic action on the soil except to contribute to the total salt content. The presence of soluble calcium will limit its solubility.

Nitrate (NO_3^-) is not commonly found in large amounts in natural waters. Small amounts can affect the use of irrigation water by supplying plants with needed nitrogen, or in some cases with more than the desired amount of this plant nutrient. Large amounts of nitrates in water may indicate contamination from natural deposits and from many other sources such as animal wastes, sewage and decomposed soil organic matter. Nitrates, however, have no significant effect on the physical properties of soil.

Boron (B) occurs in water in various anion forms. The usual range in natural waters is from 0.01 ppm to 10 ppm boron. Concentrations greater than this are known but are most often from hot springs or brines.

In the amounts that can be tolerated by plants, boron has no measurable effect on the physical properties of soil nor on soil salinity. Boron is not as readily removed from the soil as chloride or nitrate, but most of it can be removed by successive leachings.

A small amount of boron is essential for plant growth, but a concentration slightly above the optimum is toxic to plants. Some plants are more sensitive to an excess than others. Plants grown on some sandy soils which have been irrigated for several years by water exceedingly low in boron (less than 0.02 ppm) may develop boron deficiency.

EVALUATING IRRIGATION WATER

A useful evaluation of irrigation water describes its effect on soils and plant growth. This effect is primarily related to the dissolved salts in the water. Depending upon the amount and kind of salts, different soil salinity problems develop. (*See* Table 2-4 for irrigation quality classifications.)

Table 2-4

Qualitative Classification of Irrigation Waters

	Class 1 — Excellent to Good	Class 2 — Good to Injurious	Class 3 — Injurious to Unsatisfactory
EC, ds / m	Less than 1.0	1.0 – 3.0	More than 3.0
Boron, ppm	Less than 0.5	0.5 – 2.0	More than 2.0
Sodium, percent	Less than 60	60 – 75	More than 75
Chloride, me / L	Less than 5	5 – 10	More than 10

A *saline soil* contains soluble salts in such quantities that they interfere with plant growth.

A *sodic (alkali) soil* contains enough sodium attached to the clay particles to interfere with plant growth. If a sodic soil is relatively free of soluble salts, it is called a non-saline sodic soil. If, in addition to being sodic, it has sufficient soluble salts to restrict plant growth, it is called a saline-sodic soil.

Salinity hazard — One of the hazards of irrigated agriculture is the possible accumulation of soluble salts in the root zone. (*See* Figure 2-4.) Some plants can tolerate more salts than others, but all plants have a maximum tolerance. With reasonably good irrigation practices the salt content of the saturation extract of soil is 1.5 to 3 times the salt content of the irrigation water. Where ample water is used to remove excess salt from the root zone, the salt level in the saturation extract is about 1.5 times that of the irrigation water. Where water is used more sparingly, there may be 3 times as much salt in the saturation extract.

An acre-foot of water (the amount of water covering 1 acre, 1 foot deep), which is about 325,000 gallons, weighs approximately 2,720,000

Fig. 2-4. Salts accumulate with inadequate leaching and drainage. In this photo, a salt ring has formed around a citrus tree.

pounds; therefore, 1 ppm of salt in an acre-foot of water weighs 2.72 pounds. This means that 1 acre-foot of water containing only 735 ppm (EC_w = 1.15 dS / m) carries *one ton of salt!* Many growers apply 4 feet of irrigation water per year to produce a crop — or they apply 4 tons of salt on *every acre every year!* This points out the need for adequate leaching below the root zone.

With ordinary irrigation methods there is some leaching; hence, the accumulation of salts in the soil is reduced but not eliminated. Before a critical assessment of the salinity hazard of any irrigation water is made, it is necessary to know how much salt the crop can tolerate and how much leaching is needed to reduce the salt in the soil to a satisfactory level.

Tables 2-5 through 2-9 present salt tolerance for various ornamentals. Tables 2-10, 2-11, and 2-12 show the tolerance and leaching requirements estimated for numerous crops. Leaching need not be done at each irrigation but should be done at least once a year.

Growers rotating crops must provide enough leaching so that damage to the most salt-sensitive crop in the rotation will be at a minimum.

With reasonable irrigation practices, there should be no salinity problems with irrigation water with an EC_w of less than 0.75 dS / m. Increasing problems can be expected between EC_w 0.75 and 3.0 dS / m. An EC_w greater than 3.0 will cause severe problems except for a few salt-tolerant crops.

Table 2-5

Salt Tolerance of Ornamentals[1]

Low Tolerance (EC$_w$ .75 – 1.50)[2]	Moderate Tolerance (EC$_w$ 1.50 – 3.0)	High Tolerance (More Than EC$_w$ = 3.0)
Acacia longifolia (Sydney golden wattle)	*Juniperus chinensis* (Hollywood juniper)	*Araucaria heterophylla* (Norfolk Island pine)
Cotoneaster horizontalis (Rock cotoneaster)	*Melaleuca leucadendra* (Cajeput tree)	*Arctotheca calendula* (Arctotheca)
Leptospermum laevigatum (Australian tea tree)	*Raphiolepsis indica* (Indian hawthorn)	*Baccharis pilularis* (Coyote bush)
Pachysandra terminalis (Japanese spurge)	*Agave attenuta* (Thin-leaved agave)	*Coprosma repens* (Mirror plant)
Photinia fraseri (Fraser's photinia)	*Casuarina equisetifolia* (Horsetail tree)	*Cortaderia sellowiana* (Pampas grass)
Pinus halepensis (Aleppo pine)	*Hakea suaveolens* (Needle bush)	*Delosperma alba* (White iceplant)
Rhamnus alaternus (Italian buckthorn)	*Phorium tenax* (New Zealand flax)	*Drosanthemum hispidum rose* (Lavender pink iceplant)
Strelitzia regina (Bird of Paradise)	*Pittosporum phillyraeoides* (Desert willow)	*Gazania aurantiacum* (South African daisy)
Vinca minor (Dwarf running myrtle)		*Lampranthus spectabilis* (Trailing iceplant)
Limonium perezii (Sea lavender)		*Lippia canescens repens* (Lippia)
Punica granatum (Dwarf pomegranate)		*Myoporum laetum* (Myoporum — shrub)
Crassula argentea (Jade plant)		*Myoporum parvifolium* (Myoporum — ground cover)
Festuca ovina glauca (Blue fescue)		*Pittosporum crassifolium* (Evergreen pittosporum)
Juniperus scopulorum moffeti (Moffets juniper)		
Felicia aethiopica (Felicia)		

[1]Results of work done in solution culture by Dr. Roy Branson, Extension Specialist, and Richard Maire and Lyle Pyeatt, Farm Advisors, U.C. Cooperative Extension.

[2]Assumptions include the following: EC$_{culture solution}$ ÷ 2 = EC$_e$. EC$_e$ = Electrical conductivity of the soil saturation extract, representative of the more active part of the root zone. EC$_e$ × 2 = EC$_{sw}$. EC$_{sw}$ = Electrical conductivity of soil water. EC$_w$ × 3 = EC$_{sw}$, ½EC$_{sw}$ = EC$_e$, EC$_e$, = ⅔EC$_w$. All values are expressed as dS/m.

Table 2-6

Tolerance of Ornamental Shrubs and Ground Covers to Salinity in Irrigation Water[1]

Low Tolerance[2] (EC$_W$ = .75 – 1.50)[3]	Moderate Tolerance (EC$_W$ = 1.50 – 3.0)	High Tolerance (More than EC$_W$ = 3.0)
Star jasmine (*Trachelospermum jasminoides*)	Pittosporum (*P. tobira*)	Oleander (*Nerium oleander*)
Pineapple guava (*Feijoa sellowiana*)	Viburnum (*V. tinus* v. *robustum*)	Pyracantha (*P. graeberi*)
Burford holly (*Ilex cornuta Burford*)	Texas privet (*Ligustrum lucidum*)	Rosemary (*Rosmarinus lockwoodi*)
Rose (*Rosa* sp. var. *Grenoble* on Dr. Huey root)	Lantana (*L. camara*)	Dracaena (*D. endivisa*)
Algerian ivy (*Hedera canariensis*)	Boxwood (*Buxus microphylla* v. *japonica*)	Euonymus (*E. japonica* v. *grandiflora*)
Hibiscus (*H. rosa-sinensis* cv. *Brilliante*)	Xylosma (*X. senticosa*)	Natal plum (*Carissa grandiflora*)
Heavenly bamboo (*Nandina domestica*)	Arborvitae (*Thuja orientalis*)	Bougainvillea (*B. spectabilis*)
	Dodonea (*D. viscosa* v. *atropurpurea*)	
	Silverberry (*Elaegnus pungens*)	
	Spreading juniper (*Juniperus chinensis*)	
	Bottlebrush (*Callistemon viminalis*)	

[1]Source: L. Bernstein, L. E. Francois, R. A. Clark. "Salt Tolerance of Ornamental Shrubs and Ground Covers," *J. Amer. Soc. Hort. Sci.*, 1972, 97(4):550–556.

[2]Listed in decreasing order of sensitivity. EC$_W$ values shown are associated with generally satisfactory appearance and up to 25% decrease in top growth.

[3]EC$_W$ means electrical conductivity of irrigation water (in dS/m). Assumptions include the following: EC$_e$ × 2 = EC$_{sw}$, EC$_e$ = Electrical conductivity of the soil saturation extract expressed as dS/m, representative of the more active part of the root zone. EC$_{sw}$ = Electrical conductivity of soil water, EC$_w$ × 3 = EC$_{sw}$, ½EC$_{sw}$ = EC$_e$. EC$_e$ = ⅔EC$_w$.

Table 2-7

Salt Tolerance of Shrubs and Ground Covers[1]

Low Tolerance (EC_e = Less Than 3[2])	Moderate Tolerance (EC_e = 3 – 8)	High Tolerance (EC_e = 8 – 10)
Feijoa sellowiana	*Thuja orientalis aurea nana*	*Nerium oleander*
Hybrid rose (Grenoble)	*Juniperus chinensis*	*Callistemon viminalis*
Algerian ivy	*armstronge*	*Dracaena*
Burford holly	*Lantana camara*	Natal plum
	Pyracantha graberi	Prostrate rosemary
	Pittosporum tobira	Bougainvillea
	Xylosma senticosa	*Dodonea*
	Ligustrum lucidum	*Euonymus*
	Viburnum tinus robustum	
	Silverberry *(Elaeagnus)*	
	Hibiscus	
	Star jasmine	
	Japanese boxwood	
	Heavenly bamboo *(Nandina)*	

[1]Data from the U.S. Salinity Laboratory.

[2]EC_e = Electrical conductivity of the soil saturation extract expressed as dS/m, representative of the more active part of the root zone.

Table 2-8

Salt Tolerance of Turfgrasses[1]

Low Tolerance[2] (EC_e = Less Than 4[2])	Moderate Tolerance (EC_e = 4 – 10)	High Tolerance (EC_e = 8 – 15)
Kentucky bluegrass	Alta fescue	*Purcinellia distans*
Highland bentgrass	Perennial ryegrass	Common bermuda
Red fescue		Tiffway
Meadow fescue		Tiffgreen
		Sunturf
		Seaside bentgrass
		Zoysia
		St. Augustine

[1]O. R. Lunt and V. B. Youngner, *Agronomy Journal,* Vol. 53, July-August 1961, pp. 247–249.

[2]EC_e = Electrical conductivity of the soil saturation extract expressed as dS/m, representative of the more active part of the root zone.

Table 2-9

Salt Tolerance of Floriculture Crops[1]

Common Name and Cultivar	EC_e (dS/m at 25°C for 10% yield decrease[2])
Chrysanthemum	
Fresh weight response cv. Bronz Kramer	6.0
cv. Albatross	2.0
Stock	
Height response ave. 6 cv.	4.0
Rose	
Fresh weight cv. Better Times	3.5
Carnation	
Bloom yield cv. Sims	1.5
Bloom yield cv. Sims	3.0
Quality of blooms Sims	2.5
Poinsettia	
% leaf abscission cv. Barbara Ecke Supreme	2.5
Exact diameter cv. Barbara Ecke Supreme	4.0
China aster	
Fresh weight cv. King	2.0
Geranium	
Cuttings produced ave. 3 cv.	1.5
Lily	
Height cv. Croft	1.5
Gladiolus	
Spike weight ave. cv. Spotlight and Valoras	1.5
Spike length ave. cv. Spotlight and Valoras	1.5
Wt. cormels produced Spotlight and Valoras	1.0
African Violet	
Fresh weight cv. Mentor Boy	1.5
Gardenia	
Dry weight cv. Belmont	1.0
Azalea	
Dry weight cv. Sweetheart Supreme	1.0
Dry weight cv. Mrs. Fred Saunders	1.0

[1]Data based on work conducted by A. M. Kofranek, H. C. Kohl, Jr., and O. R. Lunt. Most of these papers were published in the *Proc. Amer. Soc. for Hort. Science* for the years 1952 through 1959.

[2]EC_e = Electrical conductivity of the soil saturation extract expressed as dS/m, representative of the more active part of the root zone.

Table 2-10

Vegetables — Reduction in Yield[1]

Crop	EC$_e$[2] (0%)	EC$_w$[3]	LR	EC$_e$ (10%)	EC$_w$	LR	EC$_e$ (25%)	EC$_w$	LR	EC$_e$ (50%)	EC$_w$	LR	EC$_e$[4] (Maximum)
Beets[5]	4.0	2.7	9%	5.1	3.4	11%	6.8	4.5	15%	9.6	6.4	21%	15.0
Broccoli	2.8	1.9	7%	3.9	2.6	10%	5.5	3.7	14%	8.2	5.5	20%	13.5
Tomatoes	2.5	1.7	7%	3.5	2.3	9%	5.0	3.4	14%	7.6	5.0	20%	12.5
Cantaloupes	2.2	1.5	5%	3.6	2.4	8%	5.7	3.8	12%	9.1	6.1	19%	16.0
Cucumbers	2.5	1.7	8%	3.3	2.2	11%	4.4	2.9	14%	6.3	4.2	21%	10.0
Spinach	2.0	1.3	4%	3.3	2.2	7%	5.3	3.5	12%	8.6	5.7	21%	15.0
Cabbage	1.8	1.2	5%	2.8	1.9	8%	4.4	2.9	12%	7.0	4.6	19%	12.0
Potatoes	1.7	1.1	6%	2.5	1.7	9%	3.8	2.5	12%	5.9	3.9	20%	10.0
Sweet corn	1.7	1.1	6%	2.5	1.7	9%	3.8	2.5	13%	5.9	3.9	20%	10.0
Sweet potatoes	1.5	1.0	5%	2.4	1.6	8%	3.8	2.5	13%	6.0	4.0	20%	10.5
Peppers	1.5	1.0	6%	2.2	1.5	9%	3.3	2.2	13%	5.1	3.4	19%	8.5
Lettuce	1.3	0.9	5%	2.1	1.4	8%	3.2	2.1	12%	5.2	3.4	20%	9.0
Radishes	1.2	0.8	4%	2.0	1.3	7%	3.1	2.1	12%	5.0	3.4	19%	9.0
Onions	1.2	0.8	5%	1.8	1.2	8%	2.8	1.8	12%	4.3	2.9	19%	7.5
Carrots	1.0	0.7	4%	1.7	1.1	7%	2.8	1.9	12%	4.6	3.1	19%	8.0
Beans	1.0	0.7	5%	1.5	1.0	8%	2.3	1.5	12%	3.6	2.4	18%	6.5

[1] Adapted from "Quality of Water for Irrigation." R. S. Ayers. *Jour. of the Irrig. and Drain. Div.*, ASCE. Vol. 103. No. IR2. June 1977, p. 141.

[2] EC$_e$ means electrical conductivity of the saturation extract of the soil reported in dS / m at 25°C.

[3] EC$_w$ means electrical conductivity of the irrigation water in dS / m at 25°C.

[4] Maximum EC$_e$ is the electrical conductivity of the soil saturation extract at which crop growth ceases.

[5] Sensitive during germination. EC$_e$ should not exceed 3 dS / m.

Table 2-11

Fruit and Nuts — Reduction in Yield[1]

Crop	EC_e[2]	EC_w[3]	LR	EC_e	EC_w	LR	EC_e	EC_w	LR	EC_e	EC_w	LR	EC_e[4]
	(0%)			(10%)			(25%)			(50%)			(Maximum)
Date palms	4.0	2.7	4%	6.8	4.5	7%	10.9	7.3	11%	17.9	12.0	19%	32.0
Figs }													
Olives	2.7	1.8	6%	3.8	2.6	9%	5.5	3.7	13%	8.4	5.6	20%	14.0
Pomegranates }													
Grapefruit	1.8	1.2	8%	2.4	1.6	10%	3.4	2.2	14%	4.9	3.3	21%	8.0
Oranges	1.7	1.1	7%	2.3	1.6	10%	3.2	2.2	14%	4.8	3.2	20%	8.0
Lemons	1.7	1.1	7%	2.3	1.6	10%	3.3	2.2	14%	4.8	3.2	20%	8.0
Apples }													
Pears	1.7	1.0	6%	2.3	1.6	10%	3.3	2.2	14%	4.8	3.2	20%	8.0
Walnuts	1.7	1.1	7%	2.3	1.6	10%	3.3	2.2	14%	4.8	3.2	20%	8.0
Peaches	1.7	1.1	8%	2.2	1.4	11%	2.9	1.9	15%	4.1	2.7	21%	6.5
Apricots	1.6	1.1	9%	2.0	1.3	11%	2.6	1.8	15%	3.7	2.5	21%	6.0
Grapes	1.5	1.0	4%	2.5	1.7	7%	4.1	2.7	11%	6.7	4.5	19%	12.0
Almonds	1.5	1.0	7%	2.0	1.4	10%	2.8	1.9	14%	4.1	2.7	19%	7.0
Plums	1.5	1.0	7%	2.1	1.4	10%	2.9	1.9	14%	4.3	2.8	20%	7.0
Blackberries	1.5	1.0	8%	2.0	1.3	11%	2.6	1.8	15%	3.8	2.5	21%	6.0
Boysenberries	1.5	1.0	8%	2.0	1.3	11%	2.6	1.8	15%	3.8	2.5	21%	6.0
Avocados	1.3	0.9	8%	1.8	1.2	10%	2.5	1.7	14%	3.7	2.4	20%	6.0
Raspberries	1.0	0.7	6%	1.4	1.0	9%	2.1	1.4	13%	3.2	2.1	19%	5.5
Strawberries	1.0	0.7	9%	1.3	0.9	11%	1.8	1.2	15%	2.5	1.7	21%	4.0

[1] Adapted from "Quality of Water for Irrigation." R. S. Ayers. Jour. of the Irrig. and Drain. Div., ASCE. Vol. 103, No. IR2, June 1977, p. 142.

[2] EC_e means electrical conductivity of the saturation extract of the soil reported in dS / m at 25°C.

[3] EC_w means electrical conductivity of the irrigation water in dS / m at 25°C.

[4] Maximum EC_e is the electrical conductivity of the soil saturation extract at which crop growth ceases.

Table 2-12

Grasses, Cover and Mulches—Reduction in Yield[1]

Crop	EC_e[2] (0%)	EC_w[3] (0%)	LR (0%)	EC_e (10%)	EC_w (10%)	LR (10%)	EC_e (25%)	EC_w (25%)	LR (25%)	EC_e (50%)	EC_w (50%)	LR (50%)	EC_e[4] (Maximum)
Tall wheat grass	7.5	5.0	8%	9.9	6.6	10%	13.3	9.0	14%	19.4	13.0	21%	31.5
Wheat grass (fairway)	7.5	5.0	11%	9.0	6.0	14%	11.0	7.4	17%	15.0	9.8	22%	22.0
Bermudagrass	6.9	4.6	10%	8.5	5.7	13%	10.8	7.2	16%	14.7	9.8	22%	22.5
Perennial ryegrass	5.6	3.7	10%	6.9	4.6	12%	8.9	5.9	16%	12.2	8.1	21%	19.0
Birdsfoot trefoil, narrow leaf	5.0	3.3	11%	6.0	4.0	13%	7.5	5.0	17%	10.0	6.7	22%	15.0
Harding grass	4.6	3.1	9%	5.9	3.9	11%	7.9	5.3	15%	11.1	7.4	21%	18.0
Tall fescue	3.9	2.6	6%	5.8	3.9	8%	8.6	5.7	12%	13.3	8.9	19%	23.0
Crested wheat grass	3.5	2.3	4%	6.0	4.0	7%	9.8	6.5	11%	16.0	11.0	19%	28.5
Vetch	3.0	2.0	8%	3.9	2.6	11%	5.3	3.5	15%	7.6	5.0	21%	12.0
Sudan grass	2.8	1.9	4%	5.1	3.4	7%	8.6	5.7	11%	14.4	9.6	18%	26.0
Big trefoil	2.3	1.5	10%	2.8	1.9	13%	3.6	2.4	16%	4.9	3.3	22%	7.5
Alfalfa	2.0	1.3	4%	3.4	2.2	7%	5.4	3.6	12%	8.8	5.9	19%	15.5
Clover, berseem	1.5	1.0	3%	3.2	2.1	6%	5.9	3.9	10%	10.3	6.8	18%	19.0
Orchardgrass	1.5	1.0	3%	3.1	2.1	6%	5.5	3.7	11%	9.6	6.4	18%	17.5
Meadow foxtail	1.5	1.0	4%	2.5	1.7	7%	4.1	2.7	11%	6.7	4.5	19%	12.0
Clover, alsike, ladino, red, strawberry	1.5	1.0	5%	2.3	1.6	8%	3.6	2.4	12%	5.7	3.8	19%	10.0

[1]Adapted from "Quality of Water for Irrigation." R. S. Ayers. *Jour. of the Irrig. and Drain. Div.*, ASCE. Vol. 103, No. IR2, June 1977, p. 143.

[2]EC_e means electrical conductivity of the saturation extract of the soil reported in dS / m at 25°C.

[3]EC_w means electrical conductivity of the irrigation water in dS / m at 25°C.

[4]Maximum EC_e is the electrical conductivity of the soil saturation extract at which crop growth ceases.

It has generally been assumed that the effects of saline water could be offset by increasing the amount of leaching so that the *average* salt content of the root zone would not be increased. The USDA Salinity Laboratory has demonstrated that yields of some crops are governed not by average soil salinity but primarily by the salinity of the irrigation water. Yields were reduced as the salinity of the water increased, no matter how much leaching was done. With this in mind, the practice of blending saline drain waters with low-salt irrigation waters needs to be reassessed.

Sodium or permeability hazards — In most cases the permeability of the soil to water becomes a hazard before sodium has a toxic effect on plant growth. In a few plants, notably avocados, this may not be strictly true.

As the proportion of sodium attached to the clay in the soil increases, the soil tends to disperse or "run together," bringing about reduced rates of water penetration. The sodium adsorption ratio (SAR) indicates the relative activity of sodium ions as they react with clay. From the SAR the proportion of sodium on the clay fraction of the soil can be estimated when an irrigation water has been used for a long period of time with reasonable irrigation practice.

Most laboratories will report the SAR of irrigation waters. If not, it can be simply determined by using the following equation:

$$SAR = \frac{Na}{\sqrt{(Ca + Mg)/2}}$$

The sodium (Na), calcium (Ca) and magnesium (Mg) are expressed in me / L.

A refinement of the SAR called the *Adjusted SAR* (SAR adj) is now commonly used. SAR adj includes the added effects of precipitation and solution of calcium in soils as related to $CO_3^= + HCO_3^-$ concentrations.

$$SAR\ adj = \frac{Na}{\sqrt{(Ca + Mg)/2}}\ [1 + (8.4 - pHc)]$$

SAR adj = SAR $[1 + (8.4 - pHc)]$.

The pHc is a calculated value based on total cations, Ca + Mg and $CO_3 + HCO_3$ in the water. (For tables to calculate pHc, see the Wilcox reference at the end of the chapter.)

If the SAR adj is less than 6, there should be no problems with either sodium or permeability. In the range of 6 to 9 there are increasing problems; above 9, severe problems can be expected.

SAR adj can be reduced by:

1. Increasing the calcium content of the water by adding gypsum or some other soluble calcium salt.
2. Reducing the HCO_3 in the water by adding sulfuric acid, sulfur dioxide (SO_2) or some other acidifying amendment.

SAR adj is a good index of the sodium and permeability hazard if the water passes through the soil and reaches equilibrium with it. However, in the field this is not always the case. Water with relatively high bicarbonate and low calcium and magnesium content would have a high SAR adj and hence an exceedingly slow permeability.

Irrigation water with a very low salt content may also present a water penetration problem. In this case the addition of some salt, preferably a calcium salt such as gypsum, would be helpful. There is good evidence that all irrigation waters should contain a minimum of 20 ppm of calcium (1.0 me / L) to prevent dispersion of the soil.

TOXIC CONSTITUENTS

There are several inorganic constituents found in natural waters which can be toxic to plants. Boron, chloride and sodium are the ones commonly found. Waters high in bicarbonates have been shown to induce iron deficiencies in some plants, but this is minor when compared to their role in creating permeability problems.

Boron Hazard

A small amount of boron is necessary for plant growth. To sustain an adequate supply of this plant nutrient, 0.02 ppm B or more in the irrigation water may be required. Most waters contain an adequate supply of boron, but water from a few rivers may be deficient. Some well waters and a few surface streams contain an excess of boron, thus creating a hazard by their use. Table 2-13 provides a satisfactory guide to the boron hazard in irrigation water.

Plants grown in soils high in lime may tolerate more boron than those grown in non-calcareous soils. Table 2-14 presents the relative tolerance of plants to boron.

Table 2-13

Permissible Limits of Boron for Several Classes of Irrigation Waters

Boron Class	Sensitive Plants	Semitolerant Plants	Tolerant Plants
	----------------	(ppm)	----------------
1	0.33	0.67	1.00
2	0.33 to 0.67	0.67 to 1.33	1.00 to 2.00
3	0.67 to 1.00	1.33 to 2.00	2.00 to 3.00
4	1.00 to 1.25	2.00 to 2.50	3.00 to 3.75
5	1.25	2.50	3.75

Boron can be leached from the soil, but if concentrations are high initially, a quantity of boron sufficient to cause injury may remain after the concentration of other salts is reduced to a safe level.

Table 2-14

Relative Tolerance of Plants to Boron[1]

Very sensitive *(<0.5 ppm)*	*Less sensitive* *(0.75 – 1.0 ppm)*
Lemon[2] Blackberry	Garlic Sweet potato Sunflower
Sensitive *(0.5 – 0.75 ppm)*	Bean, mung Sesame
Avocado	Lupine
Grapefruit	Strawberry
Orange	Artichoke, Jerusalem
Apricot	Bean, kidney
Peach	Bean, lima
Cherry	Peanut
Plum	*Moderately sensitive*
Persimmon	*(1.0 – 2.0 ppm)*
Fig, Kadota	Pepper, red
Grape	Pea
Walnut	Carrot
Pecan	Radish
Cowpea	Potato
Onion	Cucumber

(Continued)

Table 2-14 (Continued)

Moderately tolerant (2.0 – 4.0 ppm)	Tolerant (4.0 – 6.0 ppm)
Lettuce	Tomato
Cabbage	Parsley
Celery	Beet, red
Turnip	
Bluegrass, Kentucky	Very tolerant
Oats	(>6.0 ppm)
Corn	Asparagus (10 – 15 ppm)
Artichoke	
Tobacco	
Mustard	
Clover, sweet	
Squash	
Muskmelon	

[1]Adapted from "Salt Tolerance of Plants." E. V. Maas, in *Handbook of Plant Science.* Maximum permissible concentration in soil water without yield or vegetative growth reduction.

[2]Plants first named are considered more sensitive and last named are more tolerant.

Table 2-15

Chloride Toxicity Index[1]

Chlorides (me / L)	Chlorides (ppm)	Notes
Less than 2	Less than 70	Generally safe for all plants.
2 – 4	70 – 140	Sensitive plants usually show slight to moderate injury.
4 – 10	140 – 350	Moderately tolerant plants usually show slight to substantial injury.
Greater than 10	Greater than 350	Can cause severe problems.

[1]These values refer to chloride concentrations in the irrigation water. See Table 2-16 for soil extract values.

Chloride Hazard

Chlorides are found in all natural waters. The common chlorides are soluble and are not fixed in the soil. This allows them to move through the soil and into the drainage water. Chlorides in relatively small amounts are necessary for plant growth. In high concentrations, however, chlorides will inhibit plant growth, and they are specifically toxic to some plants.

Most annual crops and short-lived perennials are moderately to highly tolerant to chlorides. Trees, vines and woody ornamentals are sensitive to chlorides. For these plants, Table 2-16 may serve as a guide.

Sodium Toxicity

Trees, vines and woody ornamentals may be sensitive to excessive sodium absorbed through the plant roots. As in the case of chlorides, annual crops are usually not affected except for the contribution of sodium to total salt content. For waters with SAR adj below 3 there are no problems; from SAR adj values of 3 to 9, problems increase, and above 9 they are severe. For waters this high, severe soil permeability problems would also exist.

The increased use of recycled water poses potential risks to soil physical structure, which affects water penetration, and because of phytotoxicity to the plants being irrigated. Recycled water often contains high levels of sodium, boron and chloride, which can be phytotoxic at relatively low concentrations. Special attention is needed to assure that waste water does not unnecessarily degrade the areas where it is used for irrigation purposes.

WATER FOR SPRINKLER IRRIGATION

Most water suitable for surface irrigation may be safely used for overhead sprinkler irrigation. There are, however, some exceptions.

Leaf burn caused by sodium and chloride absorption may occur when the rate of evaporation is high. Conditions such as low humidity, high temperature and winds can increase the concentration of these ions in the water on the leaves between rotations of the sprinklers. Sometimes this can be corrected by increasing the rate of rotation. If this is impractical, it may be necessary to irrigate only at night during periods of hot, dry weather. Usually there is no problem when the irrigation water contains 3 me / L or less of either sodium or chloride.

Bicarbonate ions in water can also be a problem with overhead sprinkler irrigation. A white deposit of calcium carbonate forms on the leaves and fruit. This can render some fruits and ornamentals unmarketable because they are unattractive. This coat of "whitewash" is not known to have any other adverse effect on plant growth. Levels below 1.5 me / L of HCO_3 should cause no problem. Table 2-16 shows the relative tolerance of plants to chloride levels.

DRIP IRRIGATION

Drip irrigation is the frequent slow application of water through various types of emitters. Drip systems vary according to the crop. For example, a mature tree may require two to six emitters, each emitter applying water at the rate of ½ to 2 gallons per hour.

These low rates of application require small orifices and therefore need water which has been filtered free of solid particles. Dissolved salts in the water may crystallize around the orifices and reduce their size or plug them completely.

Table 2-16

Tolerance of Some Plants to Chloride Levels in Saturated Extract of Soil

Crops		Chloride[1] (me / L)
Fruit crops	*Rootstock*	
Citrus	Rangput lime, Cleopatra mandarin	25
	Rough lemon, tangelo, sour orange	15
	Sweet orange, citrange	10
Stone fruit	Mariana	25
	Lovell, Shalil	10
	Yunnan	7
Avocado	West Indian	8
	Mexican	5
	Varieties	
Grapes	Thompson seedless, Perlette	25
	Cardinal, Black Rose	10
Berries	Boysenberry	10
	Olallie blackberry	10
Strawberry	Lassen	8
	Shasta	5
Miscellaneous crops		
Beets		Near 90
Corn (young)		70
Aster		Near 50
Gladiolus		Near 40
Tomato		39
Geranium		Near 30
Gardenia		Near 25
Beans, kidney		24
Rhodesgrass		24
Dallis grass		19
Beans, navy		18

[1]This figure is usually four to six times greater than the chloride content in the irrigation water used.

The total amount of water used in drip irrigation is usually less than for conventional irrigation systems because:

1. Tailwater runoff is completely eliminated.
2. Evaporation from the soil surface is reduced because less of the soil surface is wetted.
3. The total volume of soil wetted usually is less.
4. Deep percolation of water may be reduced.

Drip irrigation is more adapted to permanent crops such as orchards and vineyards, especially on irregular slopes. Since it may save water, it is also adapted to areas where water is costly or scarce.

Drip irrigation systems are used on some annual crops where costs can be justified.

CROP SALINITY TOLERANCE AND LEACHING REQUIREMENT

The plant tolerance tables (Tables 2-10 through 2-12) show:

1. The yield reduction to be expected due to salinity of irrigation water (EC_w).
2. The salt content of the soil saturation extract (EC_e) when common surface irrigation methods are used.
3. The leaching requirement (LR), which is the fraction of the irrigation water that must be leached through the active root zone to control salinity at a specific level.
4. The maximum concentration of salts that the plant can tolerate (EC_e Max).

These tables are from data developed by the USDA Salinity Laboratory, Riverside, California.

SALT MOVEMENT IN SOIL

Soluble salts in the soil move in the direction of water movement. In areas of overall flooding or sprinkling, salt movement is directly downward. Movement and concentration of salts on beds, border checks or berms in orchards or vineyards commonly occur.

Seedlings are usually more sensitive to soluble salts in the soil than are established plants. Planting on the shoulder of the bed or planting two rows on a wide bed so that the salts will be pushed to the center of the bed and away from the seeds or seedlings may help to prevent damage (see Figure 2-5). If either the soil or the irrigation water is marginally saline, irrigating in alternate furrows may be necessary. Sprinkling irrigated seedlings until they become well established may also be desirable.

Excessive salts may accumulate in the tops of beds during pre-irrigation. This is particularly true where animal manures have been used and subsequent rainfall has not been great enough to remove excess salt before planting the seed. In this case, it may be necessary to sprinkle irrigate to remove the salt or to disperse the salts before seeding.

PATTERNS OF SALT ACCUMULATION IN THE FIELD

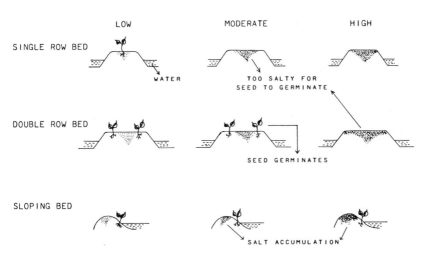

Fig. 2-5. Modification of seedbeds permits germination of seeds for good stand establishment.

In areas of low rainfall, permanent berms in the rows of orchards and vineyards may accumulate excessive salts in a few years even where relatively low-salt water is used. In such cases, these berms should be removed to disperse the salt and then be rebuilt.

Salinity is usually not a problem in modern container nurseries. But where irrigation and water quality are poor, salt accumulation may be a problem. Under these conditions we would expect accumulation to occur as pictured in Figure 2-6b. Salt accumulation may also occur where single low-volume emitters do not deliver enough water to wet the entire soil volume (see Figure 2-6c).

PATTERNS OF SALT ACCUMULATION IN NURSERY POTS

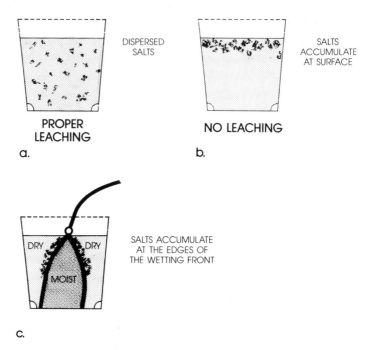

Fig. 2-6. Water management changes pattern of salt distribution in containers.

DRAINAGE

Most crops grow best where the water table is more than 6 feet below the soil surface. Fields where water stands within 6 feet of the surface should be drained. This can be accomplished by the installation of a tile or other drainage system (Figure 2-7). Tile will drain water only if placed below the water table or zone of saturation. Tile will not drain soils which are not saturated with water. If a field is poorly drained and there is an outlet for the drainage water, a carefully designed and installed drainage system may pay handsomely. These shallow waters are exceedingly saline in many areas and may not be suitable for irrigation.

If carefully managed, many crops can get a large share of their water needs from shallow ground water, provided the ground water is not excessively saline.

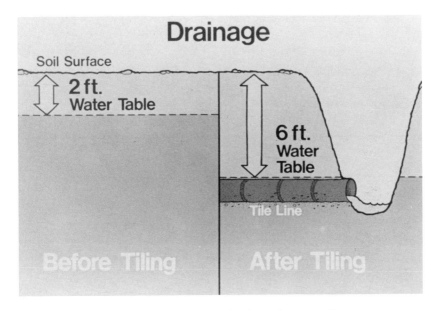

Fig. 2-7. Tile drainage is used to lower the water table.

Table 2-17

Conversion Table for Units of Flow[1]

Units	Cubic Feet per Second	Gallons per Minute	Million Gallons per Day	Acre-Inches per 24 Hours	Acre-Feet per 24 Hours
Cu. ft. per second	1.0	448.8	0.646	23.80	1.984
Gal. per minute	0.00223	1.0	0.00144	0.053	0.00442
Million gal. per day	1.547	694.4	1.0	36.84	3.07
Acre-inches per 24 hours	0.042	18.86	0.0271	1.0	0.0833
Acre-feet per 24 hours	0.504	226.3	0.3259	12.0	1.0

[1]The following approximate formulas may be conveniently used to compute the depth of water applied to a field.

$$\frac{\text{cu. ft. per second} \times \text{hours}}{\text{acres}} = \text{acre-inches per acre, or average depth in in.}$$

$$\frac{\text{gal. per minute} \times \text{hours}}{450 \times \text{acres}} = \text{acre-inches per acre, or average depth in in.}$$

Table 2-18

Units of Water Measurement

Volume units

One acre-inch

= 3,630 cu ft.
= 27,154 gal.
= $\frac{1}{12}$ acre-foot

One acre-foot

= 43,560 cu ft.
= 325,851 gal.
= 12 acre-inches

One cubic foot

= 1,728 cu in.
= 7.481 (approximately 7.5) gal.
weighs approximately 62.4 lbs. (62.5 for ordinary calculations)

One gallon

= 231 cu in.
= 0.13368 cu ft.
weighs approximately 8.33 lbs.

Flow units

One cubic foot per second

= 448.83 (approximately 450) gal. per minute
= 1 acre-inch in 1 hour and 30 seconds (approximately 1 hour), or 0.992 (approximately 1) acre-inch per hour
= 1 acre-foot in 12 hours and 6 minutes (approximately 12 hours), or 1.984 (approximately 2) acre-feet per 24 hours

One gallon per minute

= 0.00223 (approximately $\frac{1}{450}$) cu ft. per second
= 1 acre-inch in 452.6 (approximately 450) hours, or 0.00221 acre-inch per hour
= 1 acre-foot in 226.3 days, or 0.00442 acre-foot per day
= 1 in. depth of water over 96.3 sq. ft. in 1 hour

Million gallons per day

= 1.547 cu ft. per second
= 694.4 gal. per minute

SUPPLEMENTARY READING

1. *Diagnosis and Improvement of Saline and Alkali Soils.* USDA Agricultural Handbook No. 60. 1954.
2. *Modern Irrigated Soils.* D. W. James, R. J. Hanks and J. J. Jurinak. Wiley-Interscience. 1982.
3. "Quality of Water for Irrigation." R. S. Ayers. *Jour. of the Irrig. and Drain. Div.,* ASCE. 103:135 – 154. June 1977.
4. "Quality of Water for Irrigation." J. D. Rhoades. *Soil Sci.* 113:277 – 284. April 1972.
5. *Salt Tolerance of Fruit Crops.* USDA Agricultural Information Bulletin 292. 1980.
6. "Salt Tolerance of Plants." E. V. Maas in *Handbook of Plant Science.* B. R. Christie, ed. CRC Press, Inc. 1984.
7. *Salt Tolerance of Plants.* USDA Agricultural Information Bulletin 283. 1964.
8. *Salt Tolerance of Vegetable Crops in the West.* USDA Agricultural Information Bulletin 205. 1959.
9. *Tables for Calculating pHc Values of Waters.* L. V. Wilcox. USDA Salinity Lab. Mimeo. December 1966.
10. *Turfgrass Water Conservation.* Coop. Ext. Ser., Univ. of Calif. Publication 21405. 1985.
11. "Water Quality for Agriculture." R. S. Ayers and D. W. Westcott. FAO. Rome. 1976.
12. *Water Relations of Plants.* P. J. Kramer, ed. Academic Press, Inc. 1983.

CHAPTER 3

Principles of Plant Growth

Webster defines *growth* as "a growing; increase; esp., progressive development of an organism, or the like." This is well illustrated by the process of planting a seed or transplanting a young plant and initiating the process that results in the production of a mature plant. Plant growth is the process that provides people with their food supply and much of their shelter. Over a century ago, noted scientist Justus von Liebig said that plant growth is "the primary source whence man and animals derive the means of their growth and support."

Growth is fundamentally one of the attributes of living protoplasm. It is more than a mere increase in size and weight, although this may be one expression of it. Growth represents a progressive and irreversible change in form involving the formation of new cells and their enlargement and maturation into the tissues and organs of the plant.

All plants must have, in varying degrees, the same basic supply for growth of light, heat, energy, water, oxygen, carbon and mineral elements. Usually the soil supplies the needed moisture and mineral elements, while the air supplies the oxygen and carbon dioxide. Growth is stopped, started or at least modified as environmental conditions change, both in the root zone and around the aerial portion of the plants. This chapter will discuss briefly the requirements for satisfactory growth and introduce some of the concepts involved in the mineral nutrition of plants.

THE PLANT CELL

All living plant parts are made up of cells. The single plant cell is the basic structural and functional unit of the plant. It is a tiny chemical factory that absorbs and secretes materials; transforms light energy into chemical energy (photosynthesis); respires and releases energy for various activities; digests or transforms foods; synthesizes complex chemicals from air, water and simple sugars; and contains and even synthesizes the remarkable substance called protoplasm. These are but a few of the activities carried

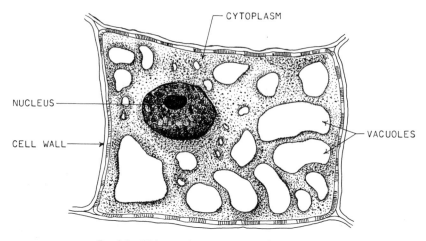

Fig. 3-1. All living plant parts are made up of cells.

on within the cell which are basic to the life process itself. A generalized plant cell is shown in Figure 3-1.

PLANT TISSUES

Groups of cells that function as a unit are referred to as *tissues*. Tissues may then be classified by function into one of four groups: meristematic, fundamental, protective and vascular. Meristematic tissues are the embryonic and undifferentiated cells, capable of growing by division and occurring at growing points. Fundamental tissues are made up of masses of cells which have little specialization in structure or function. They act primarily as storage units. The epidermal or "skin" surface of a plant usually contains protective tissues. Vascular tissues such as the xylem and the phloem function in the conductive processes of the plant. These highly specialized tissues also add mechanical support because of their structure and location.

PLANT ORGANS

The organization of a group of tissues forms an organ. Organs are generally separated into roots, leaves and stems, although throughout the plant kingdom there are plants that may have one or more of these organs missing and other specialized organs present. A careful study of botany will reveal many interesting things about plant life and contribute to one's understanding of its complexities.

ROOTS

The root is that part of the plant which ordinarily grows downward into the soil, anchoring the plant and absorbing water and mineral nutrients. It may also serve as a food storage organ or reproductive organ, or it may perform some other function. It takes many physical forms and varies in extension from a few feet to many miles on a single plant. In a careful study of a single rye plant that was allowed to grow for four months in a box 12 inches square and 22 inches deep, it was found that the accumulated length of all the roots was 387 miles. The total area of the root surface of that one plant was 6,875 square feet.

The root differs from the shoot portion of the plant primarily in structure. Unlike stems, roots do not normally bear leaves or buds and are not divided into nodes and internodes. Usually, the root differs from the shoot in function and location, but this is not always so, since some plants have roots that develop buds which give rise to leafy shoots, and other plants have aerial stems which absorb water and nutrients. Tubers and stolons are stems often found underground, while brace roots of corn and air roots of orchids, for example, are found above ground.

Root systems are often roughly grouped into two general types, fibrous and taproot. When numerous long, slender roots of about equal size are developed, they are known as fibrous roots. Examples of this root classification are corn, small grains and the grasses. If the primary root remains the largest root of the plant and continues its downward development with other roots developing from it, it is classified as a taproot. Examples of this are cotton, alfalfa, sugar beet, dandelion and ragweed. Root systems not well defined as fibrous or taproot are quite common also. Whatever the classification, it is important to know the nature of the root system of a plant if one is to know how to properly manage the growing of that plant. How extensive is the root system? How deep does it grow? How rapidly does it develop? No plant can achieve its optimum development with a poor root system. In the case of container-grown plants, if the plants are left in containers too long they will become "rootbound" (roots will fill the container). This situation can reduce potential growth and calls for replanting into larger containers or open soil.

Numerous environmental factors influence the direction and extent of root growth. These include light, gravity, temperature, salt concentration, soil texture and physical condition, oxygen supply, moisture and mineral nutrient supply. Each root becomes subject to a combination of all of these factors, and its growth is the result of the combined action of all of them.

Roots develop best in "loose" soil and in fertile soil since nitrogen and phosphorus seem to stimulate root development. Thus, one of the purposes of tillage and fertilization is to provide a proper environment for the establishment and development of a good root system.

Oxygen must be available to all living cells. The amount necessary for growth varies with species. Unless the plant has specially developed systems for transporting oxygen to the roots such as some of the swamp plants and rice, flooding of soils for appreciable lengths of time can cause root death because of lack of oxygen. This condition can be a problem with container-grown plants if adequate drainage is not provided.

Downward growth is the common response of roots to the pull of gravity. This is caused by unequal distribution of hormones inducing their effect on growth. Downward bending by the roots is known as positive geotropism, and upward bending by the shoots is negative geotropism.

The effect of light on hormonal concentration is to cause plants to bend toward the light. This again is caused by unequal hormonal distribution caused by the action of light. Bending toward light is termed positive

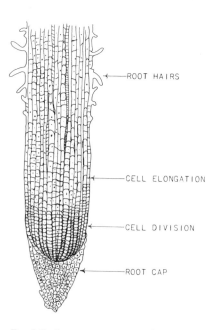

Fig. 3-2. A microscopic view of a root tip.

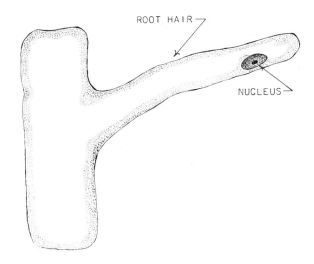

Fig. 3-3. A close look at a root-hair cell.

phototropism; bending away from light is known as negative phototropism. Some roots exhibit negative phototropism.

Roots will generally bend in the direction of most favorable temperatures, thus exhibiting a positive thermotropism, and they will grow in the direction of favorable moisture supply, positive hydrotropism. These are responses to the immediate environment and are not some sensory-perceptive seekings by the plant. Roots do not *seek* favorable temperatures or water but grow in areas of favorable temperatures or moisture to which they are intimately exposed.

A close look at a root as illustrated in Figure 3-2 will show that it is made up of many different functioning parts. The primary absorbing region for water and mineral nutrients is the younger portion near the root tip. An even closer look in this region will reveal tiny root hairs radiating outward from all sides of the root. These root hairs (Figure 3-3) are not much over 0.01 millimeter in diameter and a few millimeters in length. Each hair is an outward prolongation of a portion of an epidermal cell. The outer, delicate walls of the root hair consist partly of pectic materials, which are gelatinous and enable the root hair to cling to soil particles and to absorb water and salts in solution. It is common to find 200 to 300 root hairs per square millimeter of epidermis in the root-hair zone.

Proceeding from the growing tip back to the older root zone, the roots take on more the appearance and, to a degree, the functions of underground stems. Their primary role becomes one of conducting water and mineral nutrients from the absorbing root tip zone and transporting synthe-sized compounds from the leaves to the root tips.

A cross section of a typical root is shown in Figure 3-4. The outer layer of cells, called the epidermis, provides protection; the cortex gives sup-port, and the stele contains the conductive tissues.

As a rule, most of the water and minerals the plant obtains are taken in by the younger roots in the newly developed root area where the root hairs are most numerous. The older tissues back of this region become progres-sively impermeable, although it has been shown that there may be consid-erable water absorption through these less active regions of the root, particularly in tree crops.

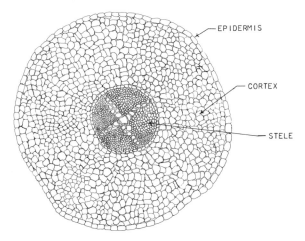

Fig. 3-4. A cross section of a typical root.

The process of absorption is one of the main functions of the root. Without a constant supply of water, the plant cannot carry on the basic physiological activities such as photosynthesis, respiration and growth. Without a supply of mineral nutrients brought in by absorptive processes, the plant would cease to live.

The study of plant physiology has brought out much concerning ab-sorption of mineral nutrients. It has been shown to be a process requiring an expenditure of energy. For example, ion concentrations may be 1,000 times greater within the cell than in the soil solution immediately outside the cell.

A brief, greatly simplified description of the absorption and move-ment of a mineral ion such as potassium will serve to illustrate the absorp-tive process. A typical young plant root cell consists of cytoplasm encircled on the outside by a cell wall (see Figure 3-1). Within the cytoplasm are vacuoles or kinds of voids holding organic solutes and inorganic ions in solution. The membrane adjoining the cell wall is termed the *plasmalemma,* while another membrane termed the *tonoplast* forms the boundary between the vacuole and the cytoplasm. It is from the outer cell wall, through the plasmalemma, across the cytoplasm and the tonoplast and into the vacuole that the potassium ion must pass. Then for it to move into the conductive xylem tissue, it must traverse other cells and pass through additional membranes.

The potassium ion diffuses from the soil to the root surface. Here it freely moves through the cell wall as it is carried in the soil solution. At this point it contacts the plasmalemma. It is this membrane which is highly impermeable to ions that control the further movement of the potassium ion into the root. Certain locations or sites are found in this membrane which are specific for specific ions. At this point it is theorized that a carrier system attaches itself to the potassium ion, transports it across the plasmalemma and deposits it on the other side. The ion is thus held inside by the membrane, and the carrier is regenerated to pick up another ion.

Once inside the cell, the potassium ion can then move by diffusion, by mass flow in the transpiration stream or by processes regulated by metabolism. Entry into the upward-moving stream focused primarily in the xylem tissues allows for rather rapid movement of the potassium ions throughout the whole plant.

Water movement is primarily a physical process. As water evaporates from leaves, it creates a difference in tension between the leaves and roots. This tension "pulls" water up the plant. Since the initial entry of water into the root is mainly through the active region of the root, into and through the cell walls, it is quite like a long tube. The major impediment to the movement is through the endodermis, where it must pass through cytoplasmic space. This is the area that is influenced by metabolism and which becomes subject to factors which affect metabolism such as temperature, oxygen supply, metabolic poisons, and carbohydrate or food supply.

Once water has entered the major conductive tissues it travels with relatively minimal resistance to flow. These tissues are the xylem and phloem. They are usually adjacent to one another and constitute the vascular tissues of the roots and the stems.

Numerous books have been written concerning subjects such as mineral nutrition of plants, plant and soil water relationships and water relations of plants. Also, hundreds of scientific articles have been written on these subjects. A few helpful references are included at the end of this chapter.

SHOOTS

The leaf and the stem make up the shoot. Each of these structures has separate but often overlapping functions. A generalized cross section of a leaf is shown in Figure 3-5. The leaf is the center of photosynthetic activity — the food manufacturing process. Its structure facilitates this function as it is physically positioned to get maximum exposure to the sun. Along its surface are minute openings called *stomates*, or *stomata*, where exchange of carbon dioxide and oxygen occurs. Underneath the surface, or

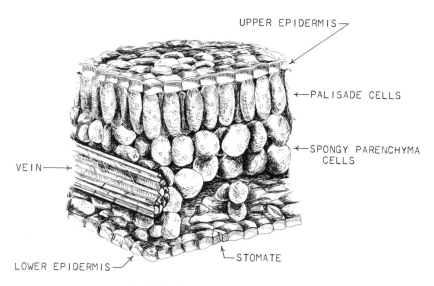

Fig. 3-5. A cross section view of a leaf.

epidermal layer, are specialized cells called the *spongy layer*, which are ideally suited to perform the photosynthetic process. Here the gases from the air, water in the cells and light from the sun all join in the presence of chloroplasts, the chlorophyll-bearing bodies, to perform the photosynthetic process.

The stems of some plants also carry on photosynthesis, although these structures usually are the conductive tissues that carry water and mineral nutrients to the leaves and photosynthesized food to the roots or other organs. They also serve to display leaves and may also become the food storage and reproductive units of the plant.

PHOTOSYNTHESIS

The manufacture of food substances from carbon dioxide and water in the presence of light and chlorophyll in living green plant tissues is *photosynthesis*. This is simply expressed in chemical equation form as:

$$6CO_2 \;+\; 12H_2O \;\xrightarrow[\text{chlorophyll}]{\text{light}}\; C_6H_{12}O_6 \;+\; 6O_2 \;+\; 6H_2O$$

carbon dioxide   water   sugar   oxygen   water

The process of respiration is just the reverse, or simply stated, the consumption of sugar in the presence of oxygen and water. In this case, light and chlorophyll are not needed.

The sugar produced by photosynthesis is transported to other plant parts. There it may be respired or stored after being converted to starch, fats, proteins and other compounds. These stored compounds provide the animal world with its basic food supply.

TRANSPIRATION

The evaporation of water from leaves, stems and other aerial parts of plants is called *transpiration*. Figure 3-6 illustrates the transpiration process. As much as 99 percent or more of the water absorbed by the roots may be transpired, yet the amount utilized is vital to the life processes of the plant. That portion evaporating has a cooling effect and serves to create a pressure pull within the conductive tissues, which helps to explain the movement of solutes and other materials in the plant's vascular system.

External factors such as light, temperature, humidity, wind velocity and soil moisture conditions influence the rate of transpiration in plants. The minute openings in the leaves called *stomata* open in response to light; thus, transpiration is much greater in the daytime than at night. Response by the stomata of opening and closure is related to changes within the living cells which are affected by external and internal factors.

Modification of plant structure is a natural process which governs plant adaptability to climatic factors. Waxy cuticles, thickened layers, decreased leaf surface, fewer stomata and hairy leaf surfaces are natural systems to modify transpiration rate and allow for greater resistance to moisture stress.

A quantity factor has been developed to express the ratio of total

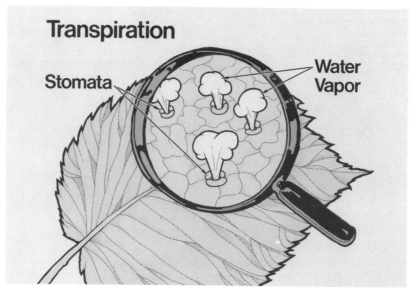

Fig. 3-6. An illustration representing the loss of water vapor through stomata in the transpiration process.

water absorbed to the total amount of dry matter produced by the plant. This is termed the "water requirement" or "transpiration ratio."

The water requirement of an adequately fertilized crop is essentially the same as one inadequately supplied with nutrients. This indicates a reduction in water required per unit of production and a better water use efficiency.

FACTORS AFFECTING GROWTH

How fast a plant grows and the shape or form it assumes are determined by internal and external factors. Heredity is the predominant internal factor, and environment is the major external factor.

Heredity

The tendency for an offspring to display the characteristics of its parents is known as *heredity.* If we consider a wheat plant, the length and strength of the stem; the shape and texture of the leaf; the number of spikelets in the head; the shape, color and surface of the glumes; the weight of the kernel; the character of the grain; the yield of the seed; and the resistance to cold, drought and disease, together with many other characteristics, have all been shown to be inheritable. The male and female sexual cells (gametes), which unite to produce the offspring, contain these specific characteristics, and the offspring is a product of that union.

The genes are the remarkable minute pieces of protoplasm located on the chromosomes of the cells that carry the genetic characteristics of the organism. These genes provide the map or the blueprint for all developing cells. Whether the plant is tall or short, the leaf is round or pointed, the flower is red or white, in short, the very genetic makeup of the plant is established by the pattern set by the combination of genes.

Mendel is known as the father of plant genetics since he set down the theory of inheritance and demonstrated that theory. Using the garden pea he was able to single out a particular characteristic, such as flower color or smoothness of the seed, and demonstrate how this characteristic was inherited. He kept accurate records of the crosses he made, knew the ancestry of each individual and found the number of individuals in the offspring that followed the characteristics of the parents. Although the science of genetics is much more complex today, the painstaking and careful scientific methods established by Mendel are still followed in genetic experimentation.

Growth vs. Time

The growth of a cell, an organ or a whole plant does not proceed at a uniform rate. Growth starts out slowly, gradually increasing until a maximum rate is reached and then slows until it ceases altogether. A characteristically S-shaped curve is obtained if total growth is plotted against time. This is illustrated in Figure 3-7.

Fluctuations in temperature, moisture supply or other environmental conditions may cause irregularities in the curve, but if the whole period of growth is considered, the shape of the curve will remain the same. When nutrient uptake is plotted against time, the accumulation of nutrients closely follows the growth curve shape. Note that it precedes the growth, since the nutrients must be present for growth to occur. A temporary shortage of nutrients will cause irregularities in the curve, and a severe shortage will essentially stop the growth. It was previously mentioned in this chapter that growth of container-grown plants may be reduced when the plants become "rootbound."

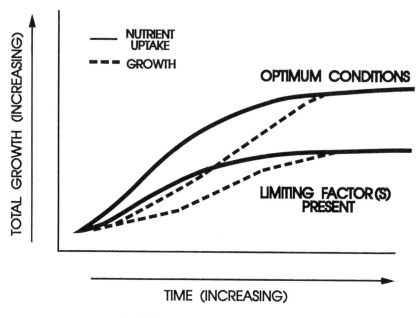

Fig. 3-7. A characteristic growth curve.

Temperature

Temperature has a marked effect upon plant growth and is one of the most important factors determining the distribution of plants over the earth's surface. The temperatures within which plants are able to grow are often designated as minimum, optimum and maximum. The points below and beyond which growth ceases are designated *minimum* and *maximum,* while the temperature at which growth proceeds best is termed *optimum.* These points are not fixed during the life of the plant, nor are they the same for all parts of the plant. In general, for temperate climates, the minimum usually falls somewhere above freezing, optimum at about 80° to 90°F and maximum around 110° to 120°F.

The direct effect of temperature must be evaluated in terms of its effect on basic processes such as photosynthesis, respiration, water and nutrient absorption, and chemical processes within the plant. Also, temperature has an effect upon the soil and chemical transformation within the soil. These are intimately related to root activity, as illustrated in reduced uptake of phosphorus from a cool soil. Microbial activity is accelerated as temperature of the soil reaches optimum and results in the release of mineral nutrient elements from soil organic matter and plant residues.

Radiant Energy

The amount, quality and duration of sunlight play an important part in plant growth and development. Most plants are able to reach maximum growth at less than full sun intensity; however, this is often modified by density of plant canopy, shading or greenhouse conditions. Some plants are better able to use the maximum sunlight. Genetic characteristics modify the growth rate.

The quality of light directly affects plant growth; for example, certain wavelengths of light trigger germination. Generally, however, light quality is not under the control of the one who grows plants. Except for certain shade-loving ornamentals, it is quite well established that the full spectrum of sunlight is generally most satisfactory for plant growth.

Photoperiodism

Photoperiodism is a term describing the behavior of a plant in relation to day length. Based upon its reaction to day length, a plant is classified as short-day, long-day or indeterminate. Short-day plants flower only under short-day conditions. If they are grown under long-day photoperiods they will not flower and will continue to grow vegetatively. Examples of short-day plants are most of the spring flowers and such autumn-flowering plants as ragweed, asters and scarlet sage.

Long-day plants attain their flowering stage only when the length of day falls within certain limits, usually 12 hours or longer. Such plants are the radish, lettuce, grains, clover and many others that normally bloom in midsummer.

Still other plants such as the tomato, cotton and buckwheat complete their reproductive cycles over a wide range of day lengths. These plants are termed *indeterminate*.

Length of day has also been reported to have an influence on the formation of tubers and bulbs; the character and extent of branching; root growth; abscission or the dropping of leaves, flowers and other plant parts; dormancy; and other effects.

Photoperiodism plays a large part in whether varieties and species may be adaptable to other areas such as moving a plant species from the north to the south and vice versa. Artificial systems can be set up to induce plants to respond contrary to what they would do if left to grow in the open. The poinsettia is forced to change its growth pattern to produce the colorful display at Christmas time by altering its photoperiod with artificial light.

In reality, the length of the dark period, not the day length, is the controlling factor in photoperiodicity. This has been established with experimentation; however, since in normal situations long days have short nights and vice versa, it is natural to assign day length to photoperiodism.

Water and Growth

The importance of water to plant growth is readily apparent. Water is required in photosynthesis, is a part of protoplasm and serves as a vehicle for translocation of food and mineral elements. The availability of water may influence the form, structure and nature of plant growth. Plants absorb more water than any other soil constituent. As previously indicated, a large part of the moisture taken up is transpired. Since water is such an important factor in plant growth and overall crop production, a separate chapter, Chapter 2, is devoted to this subject.

The Atmosphere and Growth

Quantities of nitrogen, oxygen and carbon dioxide do not vary except locally in the atmosphere. Air normally contains 78 percent nitrogen, 21 percent oxygen and 0.03 percent carbon dioxide. No higher plant is known which can make direct use of elemental nitrogen except through the action of certain microorganisms. All plants need oxygen for respiration, and the atmosphere is the most common source. Some plants are able to utilize the oxygen from oxidized compounds such as nitrates and sulfates, but this is normally confined to the microorganisms or specialized plants. All photosynthesizing plants require carbon dioxide, and this becomes the basic process by which carbon is fixed into organic matter.

Plant growth may be increased in greenhouse situations by enriching the air with carbon dioxide. This is commonly done by burning propane and exhausting the burner into the greenhouse.

The atmosphere often contains gases, particulate matter and other contaminants which can have a direct effect on plant growth. Some of these effects are positive. For example, sulfur dioxide at low levels can be absorbed by the aerial portions of some plants, and much of the nutrient need for sulfur can be satisfied this way.

Specific damage to growing plants has been observed from the pollutants in air. Products such as ozone, PAN (peroxyacetyl nitrates), excess sulfur dioxide, ethylene and fluorides come from motor vehicles, combustion of fuels, organic solvents and other sources. Most damage has been noted on fruit and vegetable crops, pine trees and flowers. Usually the injury shows up on leaves, but sometimes the plants are stunted or produce poorly. Air pollution is often increased by the relatively closed environment of a greenhouse located in highly polluted geographic areas.

Mineral Nutrient Requirements and Growth

Present information indicates the need for 16 elements in plant growth. These are carbon, hydrogen, oxygen, nitrogen, phosphorus, potassium, calcium, magnesium, sulfur, boron, chlorine, copper, iron, manganese, molybdenum and zinc. Many more elements are found in plants, but their essentiality has not been established. Some of these, listed alphabetically, are as follows: aluminum, arsenic, barium, bromine, cobalt, fluorine, iodine, lithium, nickel, selenium, silicon, sodium, strontium, titanium and vanadium. This is not a complete list, since practically all of the known elements have been isolated at one time or another from plant materials. Functions of the nutrient elements in the plant, their supply in the soil and additions in fertilizers are discussed in other chapters.

By necessity, many of the factors associated with plant growth have been omitted from this short chapter. The discussion has been simplified, and much of the basic biology and chemistry and many of the related scientific disciplines have only been touched upon. It was written to help the student, the farmer, the ornamental or turf grower or anyone interested to understand something about plant growth.

Anyone desiring to become better acquainted with the subject of plant growth is referred to the following readings.

SUPPLEMENTARY READING

1. *Biochemistry and Physiology of Plant Hormones.* T. C. Moore, Springer-Verlag. 1979.
2. *Botany: A Brief Introduction to Plant Biology,* Second Edition. T. L. Rost, M. G. Barbour, R. M. Thornton, T. E. Weir and C. R. Stocking. John Wiley & Sons, Inc. 1984.
3. *The Chemistry of Plant Processes.* C. A. Whittingham. Philosophical Library, Inc. 1965.
4. *The Climate Near the Ground,* Fourth Edition. R. Geiger. Harvard University Press. 1965.
5. *Modern Plant Biology.* H. J. Dittmer. Van Nostrand Reinhold Company. 1972.
6. *Physiology of Plant Growth and Development.* M. B. Wilkins. McGraw-Hill Book Company. 1969.

7. *Plant Growth.* M. Black and J. Edelman. Harvard University Press. 1970.
8. *Plant Physiology,* Third Edition. F. B. Salisbury and C. W. Ross. Wadsworth Publishing Co. 1985.
9. *Soil Conditions and Plant Growth,* Tenth Edition. E. W. Russell. Longman, Inc. 1974.

CHAPTER 4

Essential Plant Nutrients

The generally accepted number of essential nutrients required is 16, which includes carbon, hydrogen and oxygen, as well as 13 mineral constituents (Figure 4-1). There is evidence that at least some species may also require silicon, nickel, and cobalt either directly or indirectly. The essentiality of these for all plants has not been established.

Three of the 16 essential elements, carbon, hydrogen and oxygen, are taken primarily from the air and water. The other 13 are normally absorbed from soil by plant roots. These 13 elements are divided into three groups: primary nutrients, secondary nutrients and micronutrients. This grouping separates the elements on the basis of relative amounts required for plant growth. All of these elements are equally essential, regardless of amounts required.

CARBON, HYDROGEN AND OXYGEN

Carbon forms the skeleton for all organic molecules. Hence it is a basic building block for plant life. Carbon is taken from the atmosphere by plants in the form of carbon dioxide. Through the process of photosynthesis, carbon is combined with hydrogen and oxygen to form carbohydrates. Further chemical combinations, some with other essential elements, produce the numerous substances required for plant growth.

Oxygen is required for respiration in plant cells whereby energy is derived from the breakdown of carbohydrates. Many compounds required for plant growth processes contain oxygen. Hydrogen along with oxygen forms water, which constitutes a large proportion of the total weight of plants. Water is required for transport of minerals and plant food, and it also enters into many chemical reactions necessary for plant growth. Hydrogen is also a constituent of many other compounds necessary for plant growth. Since carbon, hydrogen and oxygen are supplied to plants primarily from the air and water, the concern over their supply is somewhat different from that of the other 13 essential elements.

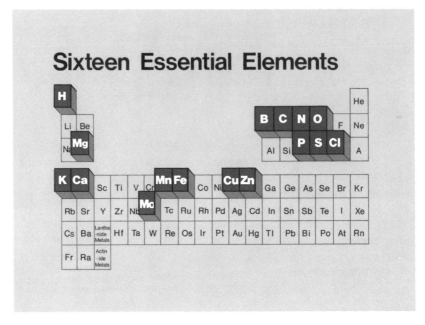

Fig. 4-1. Periodic table of elements highlighting the 16 essential plant nutrients.

PRIMARY PLANT NUTRIENTS

Nitrogen

Nitrogen is taken up by plants primarily as nitrate (NO_3^-) or ammonium (NH_4^+) ions. Plants can utilize both of these forms of nitrogen in their growth processes.

Most of the nitrogen taken up by plants is in the nitrate form. There are two basic reasons for this. First, nitrate nitrogen is mobile in the soil and moves with soil water to plant roots, where uptake can occur. Ammonical nitrogen, on the other hand, is bound to the surfaces of soil particles and cannot move to the roots. Second, all forms of nitrogen fertilizer added to soils are changed to nitrate under proper conditions of temperature, aeration, moisture, etc., by soil organisms.

Nitrogen is utilized by plants to synthesize amino acids, which in turn form proteins. The protoplasm of all living cells contains protein. Nitrogen is also required by plants for other vital compounds such as chlorophyll, nucleic acids and enzymes.

Soil Nitrogen

Most of the nitrogen in soils is unavailable to growing plants because it is tied up in organic matter. Only about 2 percent of this nitrogen becomes available each year. Since western soils generally contain relatively small amounts of organic matter, the amount of nitrogen made available each year is minor, about 0.5 pound / 1,000 sq. ft.

Many reactions involving nitrogen occur in the soil. Most of them are the result of microbial activity. Nitrogen is made available to plants from organic matter through two of these reactions. Protein and allied compounds are broken down into amino acids through a reaction called *aminization*. Soil organisms acquire energy from this digestion. They also utilize some of the amino nitrogen in their own cell structure. Ammonic nitrogen is formed by the second reaction, which converts amino compounds into ammonia (NH_3) and ammonium (NH_4^+) compounds. This reaction is called *ammonification*. The two reactions, aminization and ammonification, are referred to as *mineralization*.

Ammonical forms of nitrogen are changed to nitrate by two distinct groups of bacteria. *Nitrosomonas* and *Nitrosococcus* convert ammonia to nitrite:

$$2NH_4^+ + 3O_2 \longrightarrow 2NO_2^- + 2H_2O + 4H^+ + energy$$

ammonium nitrogen   oxygen   nitrite   water   hydrogen ions

Nitrobacter oxidizes nitrite to nitrate:

$$2NO_2^- + O_2 \longrightarrow 2NO_3^- + energy$$

nitrite   oxygen   nitrate

This two-step reaction is called *nitrification*. The reactions occur readily under conditions of warm temperature, adequate oxygen and moisture, and optimum pH.

At 75°F, nitrification may be completed in one to two weeks (Figure 4-2); at 50°F, 12 weeks or more may be required. Figure 4-3 shows the relation of soil pH to nitrification. For optimum conversion, pH must be maintained between 5.5 and 7.8.

Nitrogen may be lost from the soil to the atmosphere by reactions that convert nitrate to gaseous compounds of nitrogen. This process is called *denitrification*. Under anaerobic conditions caused by excessive moisture and / or soil compaction, certain bacteria are capable of removing oxygen

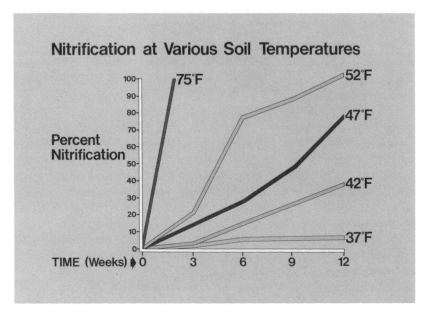

Fig. 4-2. Nitrification at various soil temperatures.

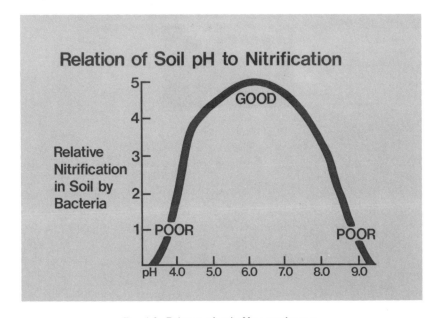

Fig. 4-3. Relation of soil pH to nitrification.

from chemical compounds in the soil to meet the needs of their life processes. When nitrate is used, various gases such as nitrous oxide (N_2O), nitric oxide (NO) and nitrogen (N_2) are formed. The reaction can be represented as follows:

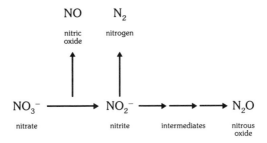

As these gases are lost from the soil into the atmosphere, there is a loss of crop-producing nitrogen from the soil.

The Nitrogen Cycle

The atmosphere contains approximately 78 percent nitrogen. It is estimated that over every 1,000 sq. ft. there are some 800 tons of nitrogen. In order for plants to utilize this nitrogen, it must be combined with hydrogen or oxygen. This process is called *nitrogen fixation*. Nitrogen may be fixed by various soil organisms. Some of these live in nodules on roots of legumes, and others are free-living organisms. Lightning also fixes smaller amounts of nitrogen which are carried into the soil by rain. The U.S. fertilizer industry fixes several million tons of nitrogen each year in various nitrogen fertilizers.

Figure 4-4 shows that nitrogen fixation by these various means provides nitrogen for growing plants. The plants in turn provide nitrogen for animals, both wild and domestic. Plant and animal wastes that are returned to the soil carry nitrogen with them.

Nitrogen may be lost to the atmosphere by denitrification or by volatilization where ammonia is produced at or near the soil surface. Leaching of nitrates, primarily from light soils, may move nitrogen below the root zone where it cannot be utilized by plants. Erosion of surface soil may also carry nitrogen from fields into streams and lakes or to the oceans. This continuous recycling of nitrogen is called the *nitrogen cycle*.

Symptoms of nitrogen deficiency in plants include:

1. Slow growth; stunting.

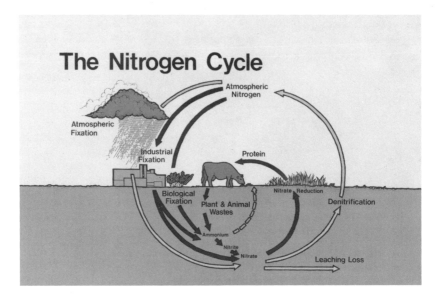

Fig. 4-4. The nitrogen cycle.

2. Yellow-green color (chlorosis).
3. "Firing" of tips and margins of leaves beginning with more mature leaves.

Chlorosis is usually more pronounced in older tissue, since nitrogen is mobile within plants and tends to move from older to younger tissue when nitrogen is in short supply.

Phosphorus

Phosphorus is absorbed by plants as $H_2PO_4^-$, $HPO_4^=$ or $PO_4^=$, depending upon soil pH. Most of the total soil phosphorus is tied up chemically in compounds of limited solubility. In neutral to alkaline soils, calcium phosphate is formed, while in acid soils, iron and aluminum phosphates are produced.

Available soil phosphorus may be only 1 percent or less of the total amount present. Solubility of phosphate is controlled by several factors. The total amount of solid-phase phosphate in the soil is one factor. The greater the total amount present in the soil, the better the chance of having more phosphorus in solution. Another important factor is the extent of contact between solid-phase phosphate and the soil solution. Greater exposure of phosphate to soil solution and to plant roots increases the ability

Plate 4-1. Nitrogen deficiency on grapes (Thompson seedless). A general fading and yellowing of leaves occurs, and shoot growth is reduced. Symptoms are difficult to recognize until deficiency becomes severe.

Plate 4-2. Nitrogen deficiency on peach. Initial yellowing of older leaves occurs. Then those leaves fall as the deficiency becomes more severe. Some chlorotic leaves may remain at terminals, as shown here.

Plate 4-3. Phosphorus deficiency on grapes. The upper row shows marginal interveinal chlorosis typical of white varieties. The leaves on younger, actively growing vines are most severely affected. The bottom row shows chlorosis and reddish speckling on a red variety. Leaf on right in each row is normal.

Plate 4.4. Phosphorus deficiency on corn. Stunted growth and purple coloration are characteristic on many plants. Symptoms usually appear during early growth, while soils are cold and root systems are small.

Plate 4-5. Potassium deficiency on squash. Spots develop on margins of more mature leaves and may coalesce to produce entire yellow or necrotic margins.

Plate 4-6. Potassium deficiency on prunes. Leaves on actively growing shoots are cupped and exhibit interveinal chlorosis. Necrotic areas develop, followed by defoliation.

Plate 4-7. Potassium deficiency on walnuts. Leaves show typical marginal chlorosis beginning with older leaves. As deficiency progresses, necrotic spots develop.

Plate 4-8. Sulfur deficiency on pear. Leaves show an overall chlorosis. Symptoms are generally more severe on young shoots.

Plate 4-9. Sulfur deficiency on tomatoes. Symptoms include stunted growth and a general yellowing of foliage. Young leaves are usually more chlorotic than mature leaves.

Plate 4-10. Magnesium deficiency on celery. Symptoms are a marginal necrosis and interveinal chlorosis beginning on older leaves, with leaves curling upward.

Plate 4-11. Magnesium deficiency on citrus (orange). Chlorosis of tip and margins of more mature leaves produces a "Christmas tree" pattern along the mid-rib of each leaf.

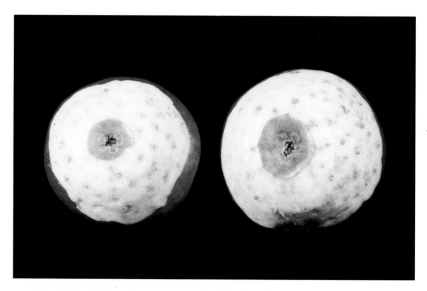

Plate 4-12. Calcium deficiency on apples. Discolored spots develop in fruit. The condition is commonly called bitter pit.

Plate 4-13. Boron deficiency on almonds. Symptoms include thickened, curled and chlorotic leaves, often with necrotic spots on margins.

Plate 4-14. Boron deficiency on grapes. Symptoms are death of shoot tips and interveinal chlorosis of leaves leading to necrosis.

Plate 4-15. Boron toxicity on grapes. Leaves show a brown speckling pattern grading into marginal necrosis.

Plate 4-16. Iron and manganese deficiencies on citrus (lemon). The center two leaves show typical iron-deficiency symptoms: green veins with yellowing of interveined areas of young leaves. The two outer leaves show characteristic manganese deficiency. Younger leaves show interveined chlorosis with gradation of pale green coloration, with darker color next to the veins.

Plate 4-17. Iron deficiency on grass. Blades exhibit a general chlorosis, turning almost white in severe cases. Grass tends to be limp and gives a matted-down appearance.

Plate 4-18. Iron deficiency on peach. Interveinal chlorosis develops on young leaves with veins as a fine network of green on a background of yellow. As deficiency progresses, leaves become entirely yellow. Symptoms often appear on poorly aerated soils.

Plate 4-19. Manganese deficiency on celery. Interveinal chlorosis appears on younger leaves first, turning necrotic as leaves mature.

Plate 4-20. Zinc deficiency on citrus (orange). Small, narrow leaves have yellow mottling between the veins. The deficiency affects younger leaves first.

Plate 4-21. Zinc deficiency on kidney beans. Interveinal chlorosis appears first on younger leavings, giving a mottled appearance. Necrotic areas may develop between veins.

Plate 4-22. Copper deficiency on citrus. Symptoms on fruit appear as scabby, reddish-brown, cracked rind. Vegetative growth may show gum blisters or pockets.

Plate 4-23. Molybdenum deficiency on cauliflower (commonly called whiptail). Symptoms include marginal scorching and cupping or rolling of leaves.

Plate 4-24. Chloride toxicity on ornamentals. Chlorosis and necrosis develop first at the tips of leaves, progressing toward the leaf base as toxicity continues.

to maintain replacement supplies. During periods of rapid growth, phosphorus in the soil solution may be replaced 10 times or more per day from solid-phase phosphorus. Soil temperature and pH also affect the solubility of phosphate. Maximum availability of soil phosphorus occurs at pH 6.5 to 7.5.

Research by several workers has demonstrated an increased uptake of phosphorus by plants when nitrogen is added with the phosphate fertilizer. This synergistic effect is especially true for banded phosphorus applications. Various explanations have been proposed for this observation. Increased root growth, physiological changes making the root cells more receptive to phosphorus, increased transfer of phosphorus across the root to the xylem, and a lowering of soil pH from ammonium nitrogen have all been suggested as reasons for the increased efficiency of phosphorus uptake in the presence of nitrogen.

Phosphorus is present in all living cells. It is utilized by plants to form nucleic acids (DNA and RNA). It is used in storage and transfer of energy through energy-rich linkages (ATP and ADP).

Phosphorus stimulates early growth and root formation. It hastens maturity and promotes seed production. Phosphorus supplementation is required most by plants under the following circumstances: (1) growth in cold weather, (2) limited root growth and (3) fast top growth. Lettuce is an example of a very highly responsive plant due to the conditions outlined. Legumes, such as peas and beans, are large users of phosphate fertilizers. Least responsive plants are trees and vines with extensive root systems raised in warm climates with long summers.

Symptoms of phosphorus deficiency in plants include:

1. Slow growth; stunting.
2. Purplish coloration on foliage of some plants.
3. Dark green coloration with tips of leaves dying.
4. Delayed maturity.
5. Poor fruit or seed development.

Potassium

Potassium is taken up by plants in the form of potassium ions (K^+). It is not synthesized into compounds, such as occurs with nitrogen and phosphorus, but tends to remain in ionic form within cells and tissues. Potassium is essential for translocation of sugars and for starch formation. It is required in the opening and closing of stomata by guard cells, which is important to efficient water use. Potassium encourages root growth and increases plant resistance to disease. It produces larger, more uniformly distributed xylem vessels throughout the root system. Potassium increases size and quality of fruit and vegetables and increases winter hardiness.

Soils may contain 900 to 14,000 pounds of potassium per 1,000 sq. ft. (6 inches deep). About 90 to 98 percent of the potassium occurs in primary minerals and is unavailable to plants. From 1 to 10 percent is trapped in expanding lattice clays and is only slowly available. Between 1 and 2 percent is contained in the soil solution and on exchange sites and is readily available to plants.

Potassium has been found to be most required by fruit trees, such as prune trees, and vegetables with a very high carbohydrate production. The most demanding vegetable has been the potato, which again is a high producer of carbohydrate as starch in the tubers. Potassium is removed from the soil in larger amounts where the entire vegetative growth is removed as is the case for asparagus, celery, cabbage and lettuce. Potassium is mobile in plant tissues.

Table 4-1

Nutrient Utilization by Various Plants[1]

Crop	Yield	N	P_2O_5	K_2O
Vegetables	----------------------	*(lbs. per 1,000 sq. ft.)*	----------------------	
Asparagus	70	2.2	1.1	2.8
Beans (snap)	230	4.0	0.9	4.6
Broccoli	410	1.8	0.7	1.7
Cabbage	1,600	6.2	1.5	5.7
Celery	3,500	6.4	3.8	17.2
Lettuce	920	2.2	0.7	4.6
Potatoes (Irish)	1,150	6.2	2.3	12.6
Squash	460	2.0	0.5	2.8
Sweet potatoes	690	3.6	1.6	7.2
Tomatoes	1,380	4.1	1.1	7.8
Fruits and nuts				
Almonds (in shell)	70	4.6	1.7	5.7
Apples	690	2.8	1.3	4.9
Cantaloupes	1,380	5.0	1.6	9.2
Grapes	690	2.9	1.0	4.5
Oranges	1,380	6.1	1.3	7.6
Peaches	690	2.2	0.9	2.8
Pears	690	2.0	0.6	2.2
Prunes	690	2.1	0.7	3.0
Turf				
Putting greens	97-120 d.m.	4.7-5.6	1.5-1.7	3.1-3.4

[1] Total uptake in harvested portion.

Symptoms of potassium deficiency in plants include:

1. Tip and marginal "burn" starting on more mature leaves.
2. Weak stalks, plants "lodge" easily.
3. Small fruit or shriveled seeds.
4. Slow growth.

SECONDARY PLANT NUTRIENTS

Calcium

Calcium is absorbed by plants as the calcium ion (Ca^{++}). A structural nutrient, it is an essential part in all walls and membranes and must be present for the formation of new cells. It is believed to counteract the toxic effects of oxalic acid by forming calcium oxalate in the vacuoles of cells. Calcium, once deposited in plant tissues, is not remobilized. Therefore, young tissue is affected first under conditions of deficiency.

Calcium is often used as a foliar spray for certain celery varieties to prevent a disorder of the young leaves called *black heart*. Further, it is common practice to use calcium sprays and / or dips on apples, pears, and cherries to reduce the fruit disorders cork spot, bitter pit, and cracking.

It is too often assumed that calcium is so generally abundant that its only other requirement as a fertilizer nutrient is on very acid soils where lime is needed. However, in alkaline soils where calcium availability can be quite low, supplemental fertilizer calcium often is required to adequately supply the plants.

Symptoms of calcium deficiency in plants include:

1. "Tip burn" of young leaves — celery, lettuce, cabbage.
2. Death of growing points (terminal buds). Root tips also affected.
3. Abnormal dark green appearance of foliage.
4. Premature shedding of blossoms and buds.
5. Weakened stems.
6. Water-soaked, discolored areas on fruits — blossom-end rot of tomatoes, peppers and melons; bitter pit or cork spot of apples and pears.

Magnesium

Plant uptake of magnesium is in the form of the magnesium ion (Mg^{++}). The chlorophyll molecule contains magnesium. It is therefore essential for photosynthesis. Magnesium serves as an activator for many plant enzymes required in growth processes.

Magnesium is mobile within plants and can be readily translocated from older to younger tissue under conditions of deficiency.

The availability of magnesium is generally high in western soils, but it is more often deficient than calcium. The most common use of magnesium fertilizer has been for celery and citrus. Crops such as these sometimes need magnesium in commercial fields to balance the generally high use of potassium from fertilizers and manure. Plants growing in sandy soils may also show deficiencies.

Symptoms of magnesium deficiency in plants include:

1. Interveinal chlorosis (yellowing) in older leaves.
2. Curling of leaves upward along margins.
3. Marginal yellowing with green "Christmas tree" area along mid-rib of leaf.

Sulfur

Uptake of sulfur by plants is in the form of sulfate ions ($SO_4^=$). Sulfur may also be absorbed from the air through leaves in areas where the atmosphere has been enriched with sulfur compounds from industrial sources.

Sulfur is a constituent of three amino acids (cystine, methionine and cysteine) and is therefore essential for protein synthesis. It is essential for nodule formation on legume roots. Sulfur is present in oil compounds responsible for the characteristic odors of plants such as garlic and onion.

Sulfur deficiencies are widespread throughout the United States and have been identified under a wide range of soil and climate conditions.

Symptoms of sulfur deficiency in plants include:

1. Young leaves light green to yellowish in color. In some plants, older tissue may be affected also.
2. Small and spindly plants.
3. Retarded growth rate and delayed maturity.

MICRONUTRIENTS

Even though micronutrients are used by plants in very small amounts, they are just as essential for plant growth as the larger amounts of primary and secondary nutrients. Care must be exercised in the use of micronutrients, since the difference between deficient and toxic levels is often small.

Micronutrients should not be applied as a "shotgun" application to cover possible deficiencies. They should be applied only when the need has been demonstrated.

Zinc

Zinc is an essential constituent of several important enzyme systems in plants. It controls the synthesis of indoleacetic acid, an important plant growth regulator. Terminal growth areas are affected first when zinc is deficient. Zinc is absorbed by plants as the zinc ion (Zn^{++}).

Zinc has been the micronutrient most often needed by western crops. Commercial citrus is generally given zinc as part of a foliar spray program one to several times a year. Many horticultural crops, including tree fruits and nuts, grapes, beans, onions and tomatoes, and some agronomic crops such as cotton, rice and corn have generally required zinc fertilization.

Symptoms of zinc deficiency in plants include:

1. Decrease in stem length and a rosetteing of terminal leaves.
2. Reduced fruit bud formation.
3. Mottled leaves (interveinal chlorosis).
4. Dieback of twigs after the first year.

Iron

Iron is required for the formation of chlorophyll in plant cells. It serves as an activator for biochemical processes such as respiration, photosynthesis and symbiotic nitrogen fixation. Iron deficiency can be induced by high levels of manganese or high lime content in soils. Iron is taken up by plants as ferrous (Fe^{++}) ions.

Iron deficiency is common in many western soils. Where deficiencies occur, it is usually due to high pH or poor aeration. Turf, certain trees and ornamentals are especially susceptible to iron deficiency.

Symptoms of iron deficiency in plants include:

1. Interveinal chlorosis of young leaves. Veins remain green except in severe cases.
2. Twig dieback.
3. In severe cases, death of entire limbs or plants.

Manganese

Manganese serves as an activator for enzymes in growth processes. It assists iron in chlorophyll formation. High manganese concentration may induce iron deficiency. Manganese uptake is primarily in the form of the manganous ion (Mn^{++}).

Manganese is generally required with zinc in foliar spraying of commercial citrus. Other tree crops may show deficiencies, but otherwise, there is no common recognition of requirements for this element in fertilizer programs in the West.

Symptoms of manganese deficiency in plants include:

1. Interveinal chlorosis of young leaves.
2. Gradation of pale green coloration with darker color next to veins. No sharp distinction between veins and interveinal areas as with iron deficiency.

Copper

Copper is an activator of several enzymes in plants. It may play a role in vitamin A production. A deficiency interferes with protein synthesis. Plant uptake is in the form of cuprous (Cu^+) and cupric (Cu^{++}) ions.

Native copper supply has been recognized as only rarely needing supplementation. These few instances have been with tree crops and plants growing on some organic and sandy soils. Copper can be highly toxic at low levels; therefore, its application is not recommended except where the need has been established.

1. Stunted growth.
2. Dieback of terminal shoots in trees.
3. Poor pigmentation.
4. Wilting and eventual death of leaf tips.
5. Formation of gum pockets around central pith in oranges.

Boron

In western soils, intensive cropping has caused boron deficiencies to become more common under a wide range of soil and climatic conditions.

Boron is taken up by plants as the borate ion ($BO_3^=$). It functions in plants in differentiation of meristem cells. With boron deficiency, cells may continue to divide, but structural components are not differentiated. Boron also apparently regulates metabolism of carbohydrates in plants. Boron, like calcium, once assimilated, is not remobilized in plants, and a continuous supply is necessary at all growing points. Deficiency is first found in the youngest tissues of the plant.

Symptoms of boron deficiency in plants include:

1. Death of terminal growth, causing lateral buds to develop and producing a "witches'-broom" effect.
2. Thickened, curled, wilted and chlorotic leaves.
3. Soft or necrotic spots in fruit or tubers.
4. Reduced flowering or improper pollination.

Although boron toxicity can occur, it is rare. Toxicities occur most often on inland desert areas associated with high-boron contaminated water.

Molybdenum

Molybdenum is taken up by plants as the molybdate ion ($MoO_4^=$). It is required by plants for utilization of nitrogen. Plants cannot transform nitrate nitrogen into amino acids without molybdenum. Legumes cannot fix atmospheric nitrogen symbiotically unless molybdenum is present.

Symptoms of molybdenum deficiency in plants include:

1. Stunting and lack of vigor. This is similar to nitrogen deficiency due to the key role of molybdenum in nitrogen utilization by plants.
2. Marginal scorching and cupping or rolling of leaves.
3. "Whiptail" of cauliflower.
4. Yellow spotting of citrus.

Chlorine

Chlorine is absorbed by plants as the chloride ion (Cl^-). It is required in photosynthetic reactions in plants. Deficiency is very rare due to its universal presence in nature.

Symptoms of chlorine deficiency in plants include:

1. Wilting, followed by chlorosis.
2. Excessive branching of lateral roots.
3. Bronzing of leaves.
4. Chlorosis and necrosis in tomatoes and barley.

NUTRIENT BALANCE

Balance is important in plant nutrition. An excess of one nutrient can cause reduced uptake of another. An excess of potassium, for example, may compete with magnesium uptake by plants. A heavy application of phosphorus may induce a zinc deficiency in soil that is marginal or low in zinc. Excess iron may induce a manganese deficiency.

Maintaining a balance of nutrients in the soil is an important management objective. By judicious use of fertilizers, nutrients which are deficient in soil can be supplied to growing plants. The objective of fertilizer programs is to supplement the capacity of soils to supply nutrients that would otherwise be deficient for normal and healthy growth.

DIAGNOSING NUTRIENT NEEDS

Diagnosing nutrient needs is covered in Chapter 10. No attempt will be made to discuss it here other than to point out one or two precautions. Visual symptoms of nutrient deficiencies can be a useful tool for diagnosing problems. Assistance should be obtained from a qualified person, since chlorosis (yellowing) and necrosis (death) of tissues can result from problems other than nutrient deficiencies. Toxicity from excessive amounts of certain elements or damage from herbicides, as well as lack of proper aeration in the root zone, can also produce yellowing or death of tissue. A trained person can usually distinguish between these various problems, whereas the average person may have difficulty. Soil and / or tissue analyses should be used to verify nutrient deficiencies and to determine their causes prior to initiating a program for correction. Testing programs are best utilized on a regular basis to predict nutrient requirements before the plants become deficient and show symptoms of stress.

SUPPLEMENTARY READING

1. *Diagnostic Criteria for Plants and Soils.* H. D. Chapman, ed. University of California, Division of Agricultural Sciences. 1966.
2. *Hunger Signs in Crops,* Third Edition. H. B. Sprague, ed. David McKay Co., Inc. 1964.
3. *Micronutrients in Agriculture.* J. J. Mortvedt, ed. Soil Sci. Soc. of America. 1972.
4. *Nitrogen in Agricultural Soils.* F. J. Stevenson, ed. American Society of Agronomy. 1982.
5. *Phosphorus for Agriculture.* Potash & Phosphate Institute. 1988.
6. *Potassium for Agriculture.* Potash & Phosphate Institute. 1987.
7. *Soil Fertility and Fertilizers,* Fourth Edition. S. L. Tisdale, W. L. Nelson and J. D. Beaton, The Macmillan Company. 1985.

CHAPTER 5

Fertilizers – A Source
of Plant Nutrients

Soil serves as a storehouse for plant nutrients and normally provides a substantial amount of the plant's nutrient requirements. Under most conditions, however, growth can be enhanced by proper application of supplemental nutrients. Any material containing one or more of the essential nutrients that are added to the soil or applied to plant foliage for the purpose of supplementing the plant nutrient supply can be called a fertilizer.

The earliest fertilizer materials were animal manures, plant and animal residues, ground bones and potash salts derived from wood ashes. Three major developments in the nineteenth century were the forerunners of the modern fertilizer industry:

1839 — The discovery of potassium salt deposits in the German states.

1842 — The treatment of ground phosphate rock with sulfuric acid to form superphosphate.

1884 — The development of the theoretical principles for combining hydrogen and atmospheric nitrogen to form ammonia.

TYPES OF FERTILIZERS

Based upon their primary nutrient content (N, P_2O_5, K_2O), fertilizers are designated as single-nutrient or multinutrient fertilizers. Single-nutrient fertilizers are called *materials* or *simple fertilizers*. Multinutrient fertilizers are referred to as *mixed fertilizers* or *complexes*.

Multinutrient fertilizers are given a numerical designation consisting of three numbers. These numbers represent respectively the nitrogen (N), phosphate (P_2O_5) and potash (K_2O) content of the fertilizer in terms of percentage by weight. This three-number designation is called a *grade*.

The fertilizer *ratio* is the relative proportion of each of the primary nu-

89

trients. For example, a 12-12-12 grade is a 1-1-1 ratio, and a 21-7-14 grade is a 3-1-2 ratio.

A zero in a grade, or ratio designation, indicates that that particular nutrient is not included in the fertilizer. For example, the grade designation for ammonium nitrate is 34-0-0, and the ratio designation for a 20-10-0 grade is 2-1-0.

NITROGEN FERTILIZERS

Anhydrous Ammonia (82-0-0)

Nitrogen from the atmosphere is the primary source of all nitrogen used by plants. This inert gas comprises about 78 percent of the earth's atmosphere. Chemical fixation of atmospheric nitrogen to form ammonia is the principal source of most nitrogen fertilizers. The production of ammonia involves the reaction of nitrogen and hydrogen in the presence of a catalyst at a high temperature and pressure.

The reaction is expressed by the following equation:

$$3H_2 + N_2 \rightleftharpoons 2NH_3$$

Gaseous ammonia is lighter than air. When compressed, it becomes a liquid about 60 percent as heavy as water. It is readily absorbed in water up to concentrations of 30 to 40 percent by weight.

The high vapor pressure of anhydrous ammonia at ordinary temperatures requires that it be transported in pressure containers.

Although the supply of nitrogen from the air is virtually limitless, sources of hydrogen are limited. In the United States, virtually all ammonia facilities use natural gas as a source of hydrogen. A ton of ammonia requires about 33,000 cubic feet of natural gas for its hydrogen requirement. Possible alternative sources of hydrogen are water, coal, oven gases and naphtha. Naphtha is frequently used in foreign plants.

Aqua Ammonia (20-0-0)

Anhydrous ammonia dissolved in water forms aqua ammonia. As usually sold, aqua ammonia contains 20 percent nitrogen, which is present in the ammonium form. Although under normal temperature conditions aqua ammonia has some free ammonia, the vapor pressure is low, making it possible to store this fertilizer in low-pressure tanks and to apply it with less expensive equipment than required for anhydrous ammonia.

To minimize loss of nitrogen, aqua (like anhydrous) ammonia should be injected below the soil's surface or below the surface of water when applied in an open irrigation system. Because of the much lower free ammonia, direct soil applications need not be made as deeply as with anhydrous ammonia.

Ammonium Nitrate (34-0-0)

Ammonium nitrate is manufactured by reacting nitric acid with anhydrous ammonia. Nitric acid is produced by the oxidation of NH_3 with air in the presence of a catalyst, usually platinum. The initial oxidation reactions are:

$$4NH_3 + 5O_2 \longrightarrow 4NO + 6H_2O$$
$$2NO + O_2 \longrightarrow 2NO_2$$

The NO_2 is then absorbed in water to give nitric acid.

$$3NO_2 + H_2O \longrightarrow 2HNO_3 + NO$$

Direct use of nitric acid is limited in the mixed fertilizer industry or for other agricultural purposes. The basic reaction for ammonium nitrate is:

$$HNO_3 + NH_3 \longrightarrow NH_4NO_3$$

The reacted material is concentrated and either prilled or spherodized. Ammonium nitrate is normally coated to prevent caking, and the commercial product carries 33.5 to 34.0 percent nitrogen.

Ammonium Sulfate (21-0-0-24S)

In the West, ammonium sulfate is a widely used fertilizer. It contains both nitrogen and sulfur. In fact, it contains more sulfur (24 percent) than nitrogen (21 percent). Ammonium sulfate is one of the oldest forms of solid nitrogen fertilizer.

Direct manufactured ammonium sulfate is made in a neutralizer-crystallizer unit by reacting anhydrous ammonia with sulfuric acid under vacuum or at atmospheric pressure.

The reaction is as follows:

$$2NH_3 + H_2SO_4 \longrightarrow (NH_4)_2SO_4$$

Crystallization is the key operation in the production of ammonium sulfate, involving two major steps: (1) the formation of nuclei in a supersaturated solution, and (2) the growth of these nuclei to the crystal size desired. Where ammonium sulfate is used in bulk blends, it is desirable to have the crystals approximately the same size as other components of the blend.

Ammonium Nitrate-Sulfate (30-0-0-6.5S)

Ammonium nitrate-sulfate is a double salt of ammonium nitrate and ammonium sulfate. It is manufactured by neutralizing a mixture of nitric

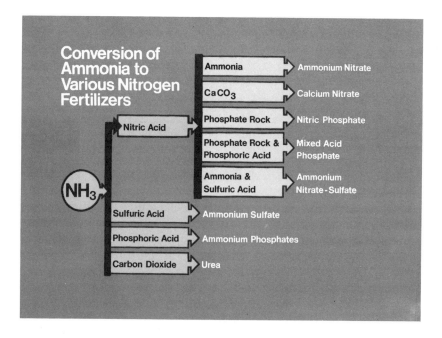

Fig. 5-1. Conversions of ammonia to various nitrogen fertilizers.

and sulfuric acids with ammonia. The nitrogen content will vary depending upon the relative percentages of ammonium nitrate and ammonium sulfate in the final product. One commercial product contains 30 percent nitrogen and 6.5 percent sulfur.

Ammonium nitrate-sulfate has good handling and storage properties.

Calcium Nitrate (15.5-0-0-19Ca)

Fertilizer grade calcium nitrate is a white, highly soluble material containing 15.5% nitrogen and 19% calcium. It is a co-product in the production of nitric phosphate fertilizers wherein crushed phosphate rock is reacted with nitric acid. The calcium nitrate containing mother liquor is neutralized with ammonia gas. Clarification, followed by successive crystallization steps, yields the double salt of calcium and ammonium nitrate. The basic reaction for calcium nitrate production is:

$$Ca^{++} + HNO_3 \xrightarrow{NH_3} 5Ca(NO_3)_2 \cdot NH_4NO_3 \cdot 10H_2O$$

(from phosphate rock)

Nitric phosphates

Nitrate of Soda (16-0-0)

This nitrogen material contains 16% nitrogen. While it is manufactured in limited quantities in this country, the bulk of this natural product is being imported from Chile. Other products have largely replaced sodium nitrate because of its relatively low nitrogen and high sodium content.

Urea (46-0-0)

Urea is a widely used nitrogen fertilizer and is also used as a protein supplement in ruminant animal feeds. The raw materials for urea are ammonia and carbon dioxide. They are reacted together in a pressure vessel at high temperature and pressure. The basic reactions are:

$$2NH_3 + CO_2 + H_2O \longrightarrow (NH_4)_2CO_3$$

ammonia   carbon dioxide   water   ammonium carbonate

$$(NH_4)_2CO_3 \longrightarrow (NH_2)_2CO + 2H_2O$$

ammonium carbonate   urea   water

The urea solution from these reactions has a concentration of about 80 percent. This solution may be used directly in urea solutions, or it may be further concentrated to make dry urea. Fertilizer grade of dry urea contains about 46 percent nitrogen.

Urea contains the highest percentage of nitrogen of commonly used solid fertilizers. Urea is very soluble and is less corrosive to equipment than many fertilizers. It is incompatible with some fertilizer materials, particularly those containing even small quantities of ammonium nitrate.

Nitrogen Solutions

Anhydrous ammonia and aqua ammonia have already been discussed. Because of the solubility of certain fertilizer salts, mainly ammonium nitrate and urea, these materials are widely used singularly or in combinations in aqueous solutions. When ammonium nitrate and urea are mixed in equal proportions their solubilities are increased, which allows for a stable solution containing more nitrogen than possible with either separate component.

Many nitrogen solutions, such as ammonium nitrate 20 percent N solution, calcium ammonium nitrate 17 percent N solution, urea – ammonium nitrate 32 percent N solution and ammonium thiosulfate 12 percent N solution, are available (see Table 5-1). Nitrogen solutions may contain ammonium nitrate and urea nitrogen, the proportion of each being dependent on the product. They are classified as pressure or non-pressure solutions. The pressure solutions are those that have an appreciable vapor pressure because of the presence of more free ammonia than the solutions can hold. Only non-pressure solutions are presented in Table 5-1.

Like any salt solution, nitrogen solutions will exhibit the phenomenon of salting out. Salting out is simply the precipitation of the dissolved salts when the temperature drops to a certain point. This point is determined by the component salts of the solution and is called the *salting out point.*

Nitrogen solutions are readily stored if the appropriate storage materials are used. Table 5-2 describes corrosion rates of various metals, alloys and other materials.

Corrosion inhibitors in nitrogen solutions reduce corrosion of carbon (mild) steel. Phosphoric acid and dichromate are commonly used inhibitors. A solution pH near 7.0 also minimizes corrosion.

PHOSPHATE FERTILIZERS

Phosphate rock deposits are the basic source of all phosphate materials. The earth in certain areas is richly endowed with natural deposits of phosphate. These deposits may be of igneous or sedimentary origin, with the latter constituting the bulk of the world's deposits. The predominant phosphate mineral in most deposits is francolite, a carbonate fluorapatite represented by the formula $Ca_{10}F_2(PO_4)_6 \cdot CaCO_3$.

Table 5-1

Composition and Physical Properties of Non-ammonia N Solutions

Total N %	8.0	10.0	11.0	12.0	17.0	20.0	20.0	23.0	28.0	30.0	32.0
NH_4NO_3 %	—	—	—	—	30.0	57.2	—	—	39.5	42.2	44.3
Urea %	—	—	—	—	—	—	43.5	50.0	30.5	32.7	35.4
$Ca(NO_3)_2$ %	—	—	—	—	44.0	—	—	—	—	—	—
Water %	—	—	—	—	26.0	42.8	56.5	50.0	30.0	25.1	20.3
Nitrate N %	—	—	—	—	11.6	10.0	—	—	7.0	7.4	7.8
Ammonium N %	8.0	10.0	11.0	12.0	5.4	10.0	20.0	23.0	7.0	7.4	7.8
Urea N %	—	—	—	—	—	—	20.0	23.0	14.0	15.2	16.4
Phosphate P_2O_5 %	24.0	34.0	37.0	—	—	—	—	—	—	—	—
Sulfur %	—	—	—	26.0	—	—	—	—	—	—	—
Spec. grav. @ 60°F	1.3	1.4	1.5	1.3	1.5	1.3	1.1	1.1	1.3	1.3	1.3
Lbs. / gal. @ 60°F	10.54	11.40	11.70	11.10	12.64	10.50	9.32	9.52	10.66	10.83	11.06
Lbs. N / gal. @ 60°F	0.84	1.16	1.29	1.33	2.15	2.10	1.86	2.18	2.98	3.25	3.54
Lbs. P_2O_5 / gal. @ 60°F	2.53	3.95	4.33	—	—	—	—	—	—	—	—
Lbs. S / gal. @ 60°F	—	—	—	2.89	—	—	—	—	—	—	—
Crystallization temp. °F	31	20	10	30	25	41	47	65	1	15	32

Table 5-2

Corrosive Characteristics of Various Metals, Alloys and Other Materials by Non-pressure Nitrogen Solutions

Not Corroded	Corroded	Materials Destroyed Rapidly
Aluminum and aluminum alloys. types 3003, 3004, 5052, 5154 and 6061	Carbon steel Cast iron	Copper Brass Bronze
Stainless steel, types 303, 304, 316, 347 and 416		Monel Zinc
Rubber		Galvanized metals
Neoprene		Usual die castings
Polyethylene, polypropylene		Concrete
Vinyl resins		
Glass, fiberglass		
Nylon		
Teflon		
Viton		
Kyrar		
Hylar		

FERTILIZERS - A SOURCE OF PLANT NUTRIENTS

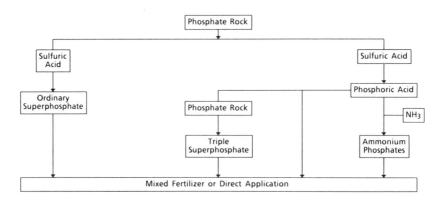

Fig. 5-2. Pathways for treating phosphate rock in the production of phosphatic fertilizers.

The principal world reserves now mined are in North Africa, North America and the USSR. The reserves in the United States are located principally in four areas: in the western states of Montana, Idaho, Utah and Wyoming; in Florida; in North Carolina; and in Tennessee.

Prior to beneficiation (refining), the phosphate matrix has a phosphate content of 14 to 35 percent. The phosphate must be concentrated for processing into fertilizers. The beneficiation process involves wet screening, hydroseparation and concentration by flotation. This product is then dried and may be calcined or ground prior to shipment. The P_2O_5 content is 31 to 33 percent.

There remains the task of converting the fluorapatite to more soluble forms that can be utilized by plants. Occasionally, however, finely ground rock phosphate is used directly on acid soils as a source of phosphorus. The pathways for treating rock phosphate used by the fertilizer industry and the end products produced are shown in Figure 5-2.

Phosphoric Acid (0-52-0) and Superphosphoric Acid (0-70-0)

Phosphoric acid is an important intermediate in the production of phosphatic fertilizers. There are two methods of producing phosphoric acid: the wet process and the furnace-grade method. The wet process is the principal method used by the fertilizer industry. Furnace grade (white acid) is used for food and industrial purposes, as well as in many specialty agricultural fertilizers. However, it is more costly to produce.

Wet-process orthophosphoric acid is produced by the action of sulfuric acid on finely ground phosphate rock. The principal chemical reaction in simplified form is:

$$Ca_{10}F_2(PO_4)_6 + 10H_2SO_4 + 20H_2O \longrightarrow 10CaSO_4 \cdot 2H_2O + 2HF$$

| phosphate rock | sulfuric acid | water | gypsum | hydrogen fluoride |

$$+ 6H_3PO_4$$

phosphoric acid

The acid is separated from the gypsum by filtration and washing to remove as much acid as possible from the gypsum cake. The acid produced contains about 30 percent P_2O_5 and is generally concentrated to the range of 40 to 54 percent P_2O_5.

Wet-process acid will frequently be green or black as a result of impurities including compounds of Fe, Al, Ca, Mg, F and organic matter. These impurities may be present in solution or solid form. Some of them may be beneficial as a source of nutrients.

Several by-products are produced during the manufacture of wet-process acid, but the principal one is gypsum. This by-product represents a disposal problem for many fertilizer manufacturers, but in areas of sodium-affected soil, by-product gypsum is used as a soil amendment. A small amount of residual phosphorus remains in the gypsum.

The trend in the fertilizer industry is to make and use higher-analysis compounds. This trend has given impetus to the development of super-phosphoric acid, which is a condensation product of orthophosphoric acid. The condensation step is illustrated in Figure 5-3.

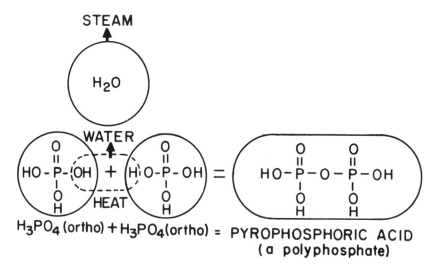

H_3PO_4 (ortho) + H_3PO_4 (ortho) = PYROPHOSPHORIC ACID (a polyphosphate)

Fig. 5-3. Removal of water from orthophosphoric acid to produce pyrophosphoric acid.

The linking of two orthophosphoric acid molecules produces pyrophosphoric acid, three molecules gives tripoly, etc., as shown in Figure 5-4.

Collectively such an acid solution is called *polyphosphoric acid* or *superphosphoric acid* and can be defined as a phosphoric acid whose molecular structure contains more than one atom of phosphorus, such as pyrophosphoric acid ($H_4P_2O_7$), tripolyphosphoric acid ($H_5P_3O_{10}$) or tetrapolyphosphoric acid ($H_6P_4O_{13}$). The relationship between the concentration of acid and the type of acid species is shown in Table 5-3.

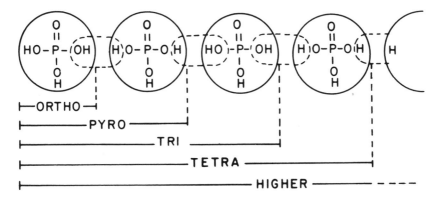

Fig. 5-4. The linkage of various members of orthophosphoric acid molecules to produce polyphosphoric acids.

Table 5-3

Forms of Phosphoric Acid at Various Concentrations[1]

		% of Total Phosphorus as				
Weight % P_2O_5	Equivalent % H_3PO_4	Ortho-phosphate	Pyro-phosphate	Tripoly-phosphate	Tetrapoly-phosphate	Higher Poly-phosphate
54	75	100				
68.8	95	100				
70	97	96	4			
72	99	90	10			
75.5	104	53	40	7		
77	106	40	47	11	2	
80	110	13	35	25	14	13
85	117	2	7	8	11	72

[1] These data are for furnace-grade acid. Wet-process acid data are generally different, and variation may occur between wet-process acid manufacturers.

Normal Superphosphate (0-20-0-12S)

Normal superphosphate is produced by reacting sulfuric acid with finely ground phosphate rock. The basic chemical reaction in simplified form is:

$$Ca_{10}F_2(PO_4)_6 + 7H_2SO_4 + 17H_2O \longrightarrow 3Ca(H_2PO_4)_2 \cdot H_2O$$

phosphate   sulfuric   water   monocalcium
rock   acid   phosphate

$$+ \; 7CaSO_4 \cdot 2H_2O \; + \; 2HF$$

gypsum   hydrogen
fluoride

The highly insoluble phosphate in the phosphate rock is converted to monocalcium phosphate monohydrate where the phosphate is approximately 85 percent water-soluble. The gypsum by-product from this reaction is intimately mixed with the monocalcium phosphate.

In the most common procedure used in the United States, sulfuric acid and phosphate rock are combined and mixed for one to two minutes. The resulting plastic mass is discharged into a compartment called a den. Retention time in the den varies from 1 to 24 hours. The acidulated phosphate sets up into a hard block and is removed from the den by various mechanical excavators. These excavators are equipped with knives to cut the block as it leaves the den. The material is stored for several weeks to allow completion of the chemical reaction. The cured product is excavated from the pile and pulverized and usually granulated before shipment. The end product usually contains 20 percent available P_2O_5.

Concentrated Superphosphate (0-45-0)

Similar procedures are involved in the production of concentrated superphosphate (sometimes called triple or treble superphosphate). The principal difference is that concentrated superphosphate is made with phosphoric acid as the acidulant, in contrast to sulfuric acid in normal superphosphate. Finely ground phosphate rock and phosphoric acid in the proper proportions are mixed together. The general reaction that occurs is:

$$Ca_{10}F_2(PO_4)_6 + 14H_3PO_4 + 10H_2O \longrightarrow 10Ca(H_2PO_4)_2 \cdot H_2O$$

phosphate   phosphoric   water   monocalcium
rock   acid   phosphate

$$+ \; 2HF$$

hydrogen
fluoride

The end product contains approximately 45 percent P_2O_5 and little $CaSO_4$.

Most superphosphates, used directly or in blends, are in granular form. Cured normal and concentrated superphosphate can be granulated through the addition of water and steam in a rotary drum granulator followed by drying and screening.

NITROGEN-PHOSPHATE COMBINATIONS

Ammonium Phosphates

The term *ammonium phosphate* encompasses a wide variety of fertilizers produced by ammoniation of phosphoric acid, often in a mixture with other materials. Such fertilizers may contain ammonium sulfate or ammonium nitrate as an additional source of nitrogen. The ammonium phosphate may be present as the diammonium or monoammonium salt, or a mixture of the two. The ammonium phosphates are widely used both in bulk blends and for direct application. Commonly used ammonium phosphates are monoammonium phosphate 11-52-0, diammonium phosphate 16-48-0 and 18-46-0 and ammonium phosphate-sulfate 16-20-0. Liquid forms of ammonium phosphates commonly used are 8-24-0, 9-30-0, 10-34-0 and 11-37-0.

Ammonium phosphates are produced by reacting ammonia with phosphoric acid in a preneutralization vessel and further ammoniating the slurry in a rotating ammoniation-granulation unit. Additional nitrogen, phosphate and sulfuric acid can be introduced into the ammoniation-granulation unit to make the desired grade. The material discharged from the unit is dried, screened and cooled before being placed in storage.

Solid ammonium polyphosphate fertilizers may be made by ammoniating superphosphoric acid. The acid is ammoniated in a water-cooled reactor at elevated temperature and pressure or in a pipe reactor at high temperature. The product is an anhydrous melt that is granulated by being mixed with cooled recycle material in a pugmill, followed by drying and cooling. The end analysis is typically 11-57-0.

Nitric Phosphates

There are three nitric phosphate processes, but all utilize the same basic reaction:

$$Ca_{10}F_2(PO_4)_6 + 20HNO_3 \longrightarrow 6H_3PO_4 + 10Ca(NO_3)_2 + 2HF$$

phosphate   nitric   phosphoric   calcium   hydrogen
rock   acid   acid   nitrate   fluoride

If the resulting solution is simply ammoniated, dried and granulated, most of the phosphate is present as dicalcium phosphate. Most nitric phosphates are produced outside North America, largely in Western Europe. The modified nitric phosphate process used in the United States employs added phosphoric acid, which results in high amounts of mono-ammonium phosphate in the final product. Since the modified nitric phosphates contain essentially equal amounts of both nitrate and ammonium forms of nitrogen, they are best described as ammonium nitrate-phosphates.

POTASH FERTILIZERS

Potassium is found throughout the world in both soluble and insoluble forms. Today only the soluble forms are economically attractive. They occur primarily as chlorides and sulfates. Potassium chloride is by far the most important source of potash. Potassium sulfate and potassium nitrate are normally used where the chloride ion may result in plant injury or in a buildup of chloride in the soil. Table 5-4 gives the properties of the principal potassium salts.

Table 5-4

Properties of Potassium Salts

		Potassium Chloride	Potassium Sulfate	Potassium Nitrate
Color		white[1]	white	white
Molecular weight		74.5	174	101
Specific gravity		1.98	2.66	2.11
Melting point		772°C	1,067°C	308°C
K_2O content, %		63	54	46.6
Solubility in water (% K_2O)	0°C	13.6	3.7	5.5
	10°C	15.0	4.6	8.1
	20°C	16.1	5.4	11.2
	30°C	17.1	6.2	14.6
	40°C	18.1	7.0	18.2
	50°C	18.9	7.7	20.5
	60°C	19.8	8.6	24.2
	70°C	20.6	8.9	27.0
	80°C	21.3	9.5	29.2
	90°C	22.2	10.0	31.1
	100°C	22.9	10.5	33.2

[1] Impurities may impart a red color to some fertilizer grades.

The greatest proportion of potash produced today is from underground deposits. These underground deposits are mined by two principal means: (1) conventional shaft mining and (2) solution mining by pumping water underground to dissolve the ore. An important consideration of commercial feasibility of a deposit is its depth, which can vary from a few hundred feet to more than 7,000 feet below the surface. Solution mining techniques allow mining depths below 4,000 feet, which is not practical with conventional shafts.

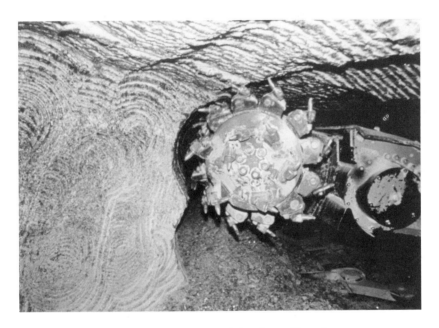

Fig. 5-5. Mining potash ore below the earth's surface.

The potash deposits in New Mexico vary from 500 to 2,500 feet below the surface. A deposit in Utah is located at about the 2,700-foot level. The Canadian beds in Saskatchewan tend to be closer to the surface in the north (3,000 feet) and go deeper as they proceed southward into the U.S. (7,000 feet).

An alternative source of potash is in natural brines in various parts of the world. Currently there are three natural brine deposits being worked in the world — Searles Lake, California, Great Salt Lake and Bonneville Lake in Utah and the Dead Sea works near Sodom, Israel.

Where conventional shaft mining processes are used, the potassium chloride (KCl) is brought to the surface as sylvinite ore. Sylvinite ore is a physical mixture of interlocked crystals of KCl and NaCl containing small quantities of dispersed clay and other impurities. The KCl is separated from the NaCl and other minerals by a selective flotation process.

SECONDARY NUTRIENTS

In the fertilizer industry and in horticulture generally, the importance of the major elements, nitrogen, phosphorus and potassium, is well established. Secondary and micronutrients are also equally recognized as essential for plant growth and for appearance. When guarantees are made for secondary and micronutrients, the usual requirement is that they be stated in terms of the elements. Table 5-5 gives the sulfur and calcium contents of most of the commonly used N, P and K fertilizers.

Calcium sources for plant nutrition are sometimes needed. Both soil-applied and foliar sprays for fruits, vegetables and various horticultural plants have included various soluble calcium sources. Other sources which provide calcium include soil amendments (lime and gypsum), manure and irrigation water.

Magnesium sources for plant nutrition include Epsom salts (magnesium sulfate), the double salt of potassium-magnesium sulfate, and magnesium nitrate used for foliar applications. On acid media, to which lime is applied, the use of dolomitic lime provides magnesium as well as calcium.

Sulfur sources for nutrition of plants are provided in many N, P and K materials. (Consult Table 5-5 for data.) Additional sulfur sources include:

Source	Percent Sulfur
Elemental sulfur	99
Gypsum	16 – 18
Sulfuric acid (95% – 99%)	32
Ferrous sulfate	11.5
Ferric sulfate	18 – 19
Calcium polysulfide solution	25
Ammonium polysulfide solution	40 – 45
Ammonium bisulfite solution (8.5% N)	17
Ammonium thiosulfate solution (12% N)	26

Other common sources of sulfur include manure, most river water, rain water and pesticidal sulfur.

Table 5-5

Average Composition of Fertilizer Materials

Fertilizer Materials	Chemical Formula	Total Nitrogen N%	Available Phosphoric Acid P_2O_5%	Water-soluble Potash K_2O%	Combined Calcium Ca%	Combined Sulfur S%	Equivalent Acidity or Basicity in Lbs. $CaCO_3$ [1] Acid	Base
Nitrogen materials								
Ammonium nitrate	NH_4NO_3	33.5 – 34					62	
Ammonium nitrate-sulfate	$NH_4NO_3•(NH_4)_2SO_4$	30				6.5	68	
Monoammonium phosphate	$NH_4H_2PO_4$	11	48				58	
Ammonium phosphate-sulfate	$NH_4H_2PO_4•(NH_4)_2SO_4$	16	20			15	88	
Ammonium polysulfide	$(NH_4)_2S_x$	20				40 – 50		
Ammonium thiosulfate	$(NH_4)_2S_2O_3$	12				26		
Diammonium phosphate	$(NH_4)_2HPO_4$	16 – 18	46 – 48				70	
Ammonium sulfate	$(NH_4)_2SO_4$	21				24	110	
Anhydrous ammonia	NH_3	82					148	
Aqua ammonia	NH_4OH	20					36	
Calcium ammonium nitrate solution	$Ca(NO_3)_2•NH_4NO_3$	17			7.6 – 8.8		9	
Calcium nitrate [2]	$Ca(NO_3)_2$	15.5			19			20
Calcium cyanamide	$CA(CN)_2$	20 – 22			37			63

(Continued)

Table 5-5 (Continued)

Fertilizer Materials	Chemical Formula	Total Nitrogen N%	Available Phosphoric Acid P_2O_5%	Water-soluble Potash K_2O%	Combined Calcium Ca%	Combined Sulfur S%	Equivalent Acidity or Basicity in Lbs. $CaCO_3$[1]	
							Acid	Base
Sodium nitrate	$NaNO_3$	16						29
Urea	$CO(NH_2)_2$	45 – 46					71	
Urea ammonium nitrate solution	$NH_4NO_3 \bullet CO(NH_2)_2$	32					57	
Phosphate materials								
Single superphosphate	$Ca(H_2PO_4)_2$		18 – 20		18 – 21	12	neutral	
Triple superphosphate	$Ca(H_2PO_4)_2$		45 – 46		12 – 14	1	neutral	
Phosphoric acid	H_3PO_4		52 – 54				110	
Superphosphoric acid	—³		76 – 83				160	
Potash materials								
Potassium chloride	KCl			60 – 62			neutral	
Potassium nitrate	KNO_3	13 – 14		44 – 46				26
Potassium sulfate	K_2SO_4			50 – 53		18	neutral	
Sulfate of potash-magnesia	$K_2SO_4 \bullet 2MgSO_4$			22	0.1	22	neutral	

[1]Equivalent per 100 lbs. of each material.

[2]Fertilizer grade calcium nitrate contains some ammonium-N.

[3]H_3PO_4, $H_4P_2O_7$, $H_5P_3O_{10}$, $H_6P_4O_{13}$ and other higher phosphate forms.

MICRONUTRIENTS

Properties of the various micronutrient sources vary considerably. A micronutrient material may be completely water-soluble or only slightly soluble. Inorganic sources may be relatively pure compounds or mixtures of compounds containing one or more micronutrients, with or without nonmicronutrient compounds. Organic sources are available as synthetic chelates or natural organic complexes of metal ions. Therefore, one classification of micronutrient sources includes (1) inorganic salts, (2) synthetic chelates and (3) natural organic complexes.

Inorganic Salts

Sulfates of Cu, Fe, Mn and Zn plus borates and molybdates are the most common sources of inorganic micronutrients. The most commonly used boron source is sodium tetraborate. This compound is relatively water-soluble. Soluble boric acid and sodium octaborate (polybor) are often used as foliar sprays.

Water-soluble ammonium and sodium molybdate are the primary sources of molybdenum. Some molybdic oxide, a slightly soluble compound, is occasionally used.

See Table 5-6 for composition and properties of inorganic micronutrient materials.

Polyphosphates will complex some metal micronutrients. However, they are not commercially produced for this purpose.

Synthetic Chelates

A chelating agent is a compound (usually organic) which can combine with a metal ion and form a ring structure between a portion of the chelating agent molecule and the metal. This delays precipitation of the metal ions in the soil as insoluble compounds. Commercially available synthetic chelating agents and the concentration of micronutrients are shown in Table 5-7. The general stability sequence of these chelates of micronutrients and calcium decreases as follows: $Fe^{+++} > Cu^{++} > Zn^{++} > Fe^{++} > Mn^{++} > Ca^{++} \geq Mg^{++}$.

Other factors may determine the relative quantities present in a given set of conditions. The relative stability of ferric chelates decreases in the order of EDDHA > DTPA > EDTA > HEEDTA > NTA, whereas the stability of these chelates of the other ions listed above decreases as follows: DTPA > EDTA > HEEDTA > NTA > EDDHA. In other words, EDDHA chelation of iron is greatly favored over the other micronutrients.

Table 5-6

Inorganic Sources of Micronutrients

Material	Element	Water Solubility	°F
	(%)	*(g / 100 g H₂O)*	
Sources of boron			
Granular borax — $Na_2B_4O_7 \cdot 10H_2O$	11.3	2.5	33
Sodium tetraborate, anhydrous — $Na_2B_4O_7$	21.5	1.3	32
Solubor® — $Na_2B_8O_{13} \cdot 4H_2O$	20.5	22	86
Ammonium pentaborate — $NH_4B_5O_8 \cdot 4H_2O$	19.9	7	64
Sources of copper			
Copper sulfate— $CuSO_4 \cdot 5H_2O$	25.0	24	32
Cuprous oxide— Cu_2O	88.8	i[1]	
Cupric oxide— CuO	79.8	i[1]	
Cuprous chloride— Cu_2Cl_2	64.2	1.5	77
Cupric chloride— $CuCl_2$	47.2	71	32
Sources of iron			
Ferrous sulfate — $FeSO_4 \cdot H_2O$	31.4	80	60
Ferrous sulfate — $FeSO_4 \cdot 7H_2O$	20.1	33	32
Ferric sulfate — $Fe_2(SO_4)_3 \cdot 9H_2O$	19.9	440	68
Iron oxalate — $Fe_2(C_2O_4)_3$	30.0	very soluble	
Ferrous ammonium sulfate —			
$Fe(NH_4)_2(SO_4)_2 \cdot 6H_2O$	14.2	18	32
Ferric chloride — $FeCl_3$	34.4	74	32
Sources of manganese			
Manganous sulfate — $MnSO_4 \cdot 4H_2O$	24.6	105	32
Manganous carbonate — $MnCO_3$	47.8	0.0065	77
Manganese oxide — Mn_3O_4	72.0	i[1]	
Manganous chloride — $MnCl_2$	43.7	63	32
Manganous oxide — MnO	77.4		
Sources of molybdenum			
Sodium molybdate — $Na_2MoO_4 \cdot H_2O$	39.7	56	32
Ammonium molybdate —			
$(NH_4)_6Mo_7O_{24} \cdot 4H_2O$	54.3	44	77
Molybdic oxide — MoO_3	66.0	0.11	64
Sources of zinc			
Zinc sulfate — $ZnSO_4 \cdot H_2O$	36.4	89	212
Zinc oxide — ZnO	80.3	i[1]	
Zinc carbonate — $ZnCO_3$	52.1	0.001	60
Zinc chloride — $ZnCl_2$	48.0	432	77
Zinc oxysulfate — $ZnO \cdot ZnSO_4$	53.8	—	—
Zinc ammonium sulfate —			
$ZnSO_4 \cdot (NH_4)_2SO_4 \cdot 6H_2O$	16.3	9.6	32
Zinc nitrate — $Zn(NO_3)_2 \cdot 6H_2O$	22.0	324	68

[1]An "i" denotes insolubility.

Table 5-7

Synthetic Chelates

Chelating Agent	Micronutrient Content, Percent Element			
	Cu	Fe	Mn	Zn
EDTA	7 – 13	5 – 14	5 – 12	6 – 14
HEEDTA	4 – 9	5 – 9	5 – 9	9
NTA	—	8	—	13
DTPA	—	10	—	—
EDDHA	—	6	—	—

Generally, the stability of metal chelates is greater near neutral than at low or high pH values. This is an important consideration when incorporating metal chelates into macronutrient fertilizers. Mixing ZnEDTA with phosphoric acid prior to ammoniation results in a breakdown of the chelate, but adding ZnEDTA with the ammoniating solution maintains a stable chelate.

Metallic salts of simple organics, such as citric acid, are also used to supply micronutrients.

Natural Organic Complexes

Many naturally occurring compounds contain chemically reactive groups similar to synthetic chelating agents. Those used commercially to complex micronutrients are often prepared from by-products of the wood pulp industry. Metal complexes of these compounds have lower stability than the common synthetic chelates. Also, they are more readily broken down by microorganisms in soil. Most are suitable for foliar sprays and mixing with fluid fertilizers.

ORGANIC PRODUCTS

Organic products and organic materials can be classified in several different ways. Strictly speaking, the term *organic* denotes carbon, including that of synthetic origin. However, organic fertilizers are usually considered naturally occurring compounds. In this category fall the wastes from sewage plants and manures. Generally they are good soil amendments. They add quantities of organic matter to the soil along with small amounts of plant nutrients. The user should be aware that sewage and industrial wastes may be contaminated with high levels of toxic elements such as cadmium and lead. Continuous use could cause excessive levels of these toxic elements to enter the plant. Table 8-2 shows the analysis of some of the most common organic products and manures.

Composts of household residues can be disposed of beneficially in improvement of the homeowner's garden. The nutrients contained in the compost can be utilized, and the organic matter will help improve soil structure. Organic materials should not be wasted, but treated as resources. It is, however, economically unsound to expect many commercial growers to meet their crops' nutrient needs through organic products.

SPECIALTY FERTILIZERS

Considerable effort has been directed toward developing fertilizers to overcome a specific problem, or to fulfill the nutrient needs for special purposes. These products are termed *specialty fertilizers.*

Nitrogen, the most widely used fertilizer element, and the most susceptible to loss by either leaching or denitrification, has received the greatest attention. Controlling the release of nitrogen can be accomplished by adding a physical barrier (coating) to water-soluble materials, using materials of limited water solubility or using materials of limited water solubility which during chemical and / or microbiological decomposition release nutrients in available forms.

Table 5-8

Analyses of Some Specialty Fertilizer Compounds

Material	Method of Controlling Release	Nutrient Content
		(%)
Resin / plastic – coated NPKS	Osmotic barrier	Varies with substrate
Sulfur-coated urea (SCU)	Coating with sulfur	32 – 37 N
Isobutylidene-diurea (IBDU)	Solubility / particle size	31 N
Urea formaldehyde (UF)	Solubility / particle size	35 – 38 N
Methylol urea (MU)	Solubility / bacterial	28 – 41 N

Coated Fertilizers

Urea, because of its high analysis and physical properties, is the principal nitrogen source used in coated fertilizers. Typically, sulfur is used as a coating to slow the release of nitrogen. Other commercial, organic coating processes for fertilizer use resin or thermoplastic coatings.

Uncoated Organic Compounds

The specialty fertilizer market is the major user of uncoated, slow-release organic nitrogen compounds. Most of these products are based on urea, since it reacts with a number of aldehydes to form compounds which are sparingly soluble in water. The most common are the ureaforms (UF), methylene ureas (MU) and isobutylidene-diurea (IBDU). In the production of the UF and MU forms, urea reacts with formaldehyde in the presence of a catalyst to form a mixture of polymers of varying chain length. Such mixtures are referred to by their generic names, *ureaforms* and *methylene ureas*. Their composition is controlled by the mole ratio of urea to formaldehyde. The rate of release is determined by the balance between the relatively soluble short-chain and the more insoluble long-chain polymers. Bacterial degradation is a major factor in subsequent nitrogen release from these products.

Unlike the UF and MU polymer mixtures, IBDU is a distinct chemical compound produced in the reaction between urea and isobutylaldehyde. It is a white compound with low water solubility. Nitrogen release is largely a function of solubility and particle size.

NUTRIENT CONVERSION FACTORS

With the exceptions of phosphorus and potassium, the nutrients in fertilizers are reported in elemental form. Conversion of phosphorus and potassium to the oxide forms or the reverse is done as follows:

$$P \times 2.29 = P_2O_5$$
$$P_2O_5 \times 0.43 = P$$
$$K \times 1.2 = K_2O$$
$$K_2O \times 0.83 = K$$

Several foreign countries report the nutrients in fertilizers on the elemental basis. Most of the scientific literature is also reported this way.

Table 5-9

Conversion Factors for Reagent Grade Materials
(fertilizer grades will differ)

To find the equivalent of one material, A, in terms of another, B, multiply the amount of A by the factor in column "A to B." To find the equivalent of material B in terms of A, multiply the amount of B by the factor in column "B to A."

A	B	Multiply A to B	B to A
Ammonia (NH_3)	Nitrogen (N)	0.8224	1.2159
Nitrate (NO_3)	Nitrogen (N)	0.2259	4.4266
Protein (crude)	Nitrogen (N)	0.1600	6.2500
Ammonium nitrate (NH_4NO_3)	Nitrogen (N)	0.3500	2.8572
Ammonium sulfate [$(NH_4)_2SO_4$]	Nitrogen (N)	0.2120	4.7168
Calcium nitrate [$Ca(NO_3)_2$]	Nitrogen (N)	0.1707	5.8572
Potassium nitrate (KNO_3)	Nitrogen (N)	0.1386	7.2176
Sodium nitrate ($NaNO_3$)	Nitrogen (N)	0.1648	6.0679
Monoammonium phosphate ($NH_4H_2PO_4$)	Nitrogen (N)	0.1218	8.2118
Diammonium phosphate [$(NH_4)_2HPO_4$]	Nitrogen (N)	0.2121	4.7138
Urea [$(NH_2)_2CO$]	Nitrogen (N)	0.4665	2.1437
Phosphoric acid (P_2O_5)[1]	Phosphorus (P)	0.4364	2.2914
Phosphate (PO_4)	Phosphorus (P)	0.3261	3.0662
Monoammonium phosphate ($NH_4H_2PO_4$)	Phosphoric acid (P_2O_5)[1]	0.6170	1.6207
Diammonium phosphate [$(NH_4)_2HPO_4$]	Phosphoric acid (P_2O_5)[1]	0.5374	1.8607
Monocalcium phosphate [$Ca(H_2PO_4)_2$]	Phosphoric acid (P_2O_5)[1]	0.6068	1.6479
Dicalcium phosphate ($CaHPO_4·2H_2O$)	Phosphoric acid (P_2O_5)[1]	0.4124	2.4247
Tricalcium phosphate [$Ca_3(PO_4)_2$]	Phosphoric acid (P_2O_5)[1]	0.4581	2.1829
Potash (K_2O)	Potassium (K)	0.8301	1.2046
Muriate of potash (KCl)	Potash (K_2O)	0.6317	1.5828
Sulfate of potash (K_2SO_4)	Potash (K_2O)	0.5405	1.8499
Potassium nitrate (KNO_3)	Potash (K_2O)	0.4658	2.1466
Potassium carbonate (K_2CO_3)	Potash (K_2O)	0.6816	1.4672
Gypsum ($CaSO_4·2H_2O$)	Calcium sulfate ($CaSO_4$)	0.7907	1.2647
Gypsum ($CaSO_4·2H_2O$)	Calcium (Ca)	0.2326	4.3000
Gypsum ($CaSO_4·2H_2O$)	Calcium oxide (CaO)	0.3257	3.0702
Calcium oxide (CaO)	Calcium (Ca)	0.7147	1.3992
Calcium carbonate ($CaCO_3$)	Calcium (Ca)	0.4004	2.4973
Calcium carbonate ($CaCO_3$)	Calcium oxide (CaO)	0.5604	1.7848
Calcium carbonate ($CaCO_3$)	Calcium hydroxide [$Ca(OH)_2$]	0.7403	1.3508
Calcium hydroxide [$Ca(OH)_2$]	Calcium (Ca)	0.5409	1.8487

(Continued)

Table 5-9 (Continued)

A	B	Multiply A to B	B to A
Magnesium oxide (MgO)	Magnesium (Mg)	0.6032	1.6579
Magnesium sulfate (MgSO$_4$)	Magnesium (Mg)	0.2020	4.9501
Epsom salts (MgSO$_4$•7H$_2$O)	Magnesium (Mg)	0.0987	10.1356
Sulfate (SO$_4$)	Sulfur (S)	0.3333	3.0000
Ammonium sulfate [(NH$_4$)$_2$SO$_4$]	Sulfur (S)	0.2426	4.1211
Gypsum (CaSO$_4$•2H$_2$O)	Sulfur (S)	0.1860	5.3750
Magnesium sulfate (MgSO$_4$)	Sulfur (S)	0.3190	3.1350
Potassium sulfate (K$_2$SO$_4$)	Sulfur (S)	0.1837	5.4438
Sulfuric acid (H$_2$SO$_4$)	Sulfur (S)	0.3269	3.0587
Borax (Na$_2$B$_4$O$_7$•10H$_2$O)	Boron (B)	0.1134	8.8129
Boron trioxide (B$_2$O$_3$)	Boron (B)	0.3107	3.2181
Sodium tetraborate pentahydrate (Na$_2$B$_4$O$_7$•5H$_2$O)	Boron (B)	0.1485	6.7315
Sodium tetraborate anhydrous (Na$_2$B$_4$O$_7$)	Boron (B)	0.2150	4.6502
Cobalt nitrate [Co(NO$_3$)$_2$•6H$_2$O]	Cobalt (Co)	0.2025	4.9383
Cobalt sulfate (CoSO$_4$•7H$_2$O)	Cobalt (Co)	0.2097	4.7690
Cobalt sulfate (CoSO$_4$)	Cobalt (Co)	0.3802	2.6299
Copper sulfate (CuSO$_4$)	Copper (Cu)	0.3981	2.5119
Copper sulfate (CuSO$_4$•5H$_2$O)	Copper (Cu)	0.2545	3.9293
Ferric sulfate [Fe$_2$(SO$_4$)$_3$]	Iron (Fe)	0.2793	3.5804
Ferrous sulfate (FeSO$_4$)	Iron (Fe)	0.3676	2.7203
Ferrous sulfate (FeSO$_4$•7H$_2$O)	Iron (Fe)	0.2009	4.9776
Manganese sulfate (MnSO$_4$)	Manganese (Mn)	0.3638	2.7486
Manganese sulfate (MnSO$_4$•4H$_2$O)	Manganese (Mn)	0.2463	4.0602
Sodium molybdate (Na$_2$MoO$_4$•2H$_2$O)	Molybdenum (Mo)	0.3965	2.5218
Sodium nitrate (NaNO$_3$)	Sodium (Na)	0.2705	3.6970
Sodium chloride (NaCl)	Sodium (Na)	0.3934	2.5417
Zinc oxide (ZnO)	Zinc (Zn)	0.8034	1.2447
Zinc sulfate (ZnSO$_4$)	Zinc (Zn)	0.4050	2.4693
Zinc sulfate (ZnSO$_4$•H$_2$O)	Zinc (Zn)	0.3643	2.7449

[1]Also called phosphoric acid anhydride, phosphorus pentoxide, available phosphoric acid.

Table 5-10

Product Density and Critical Humidity

Product	Typical Analysis	Density[1] (lbs. / cu ft.)	Critical Humidity[2] (%)
Ammonium nitrate, prilled	34-0-0	52	59
Ammonium nitrate, granulated		62	
Ammonium sulfate	21-0-0	64-68	79
Calcium nitrate	15.5-0-0	68-70	47
Urea, prilled	46-0-0	46	72
Urea, granulated		45-48	
Normal superphosphate	0-20-0	68-73	94
Triple superphosphate	0-46-0	62-70	92
Ammonium phosphate-sulfate	16-20-0	61-65	72
Monoammonium phosphate	11-48-0	56-67	92
Monoammonium phosphate	11-52-0	61	
Diammonium phosphate	18-46-0	58-67	82
Muriate of potash	0-0-60	61	84
Muriate of potash, standard		67-75	
Muriate of potash, granulated		62-68	
Muriate of potash, coarse		64-69	
Sulfate of potash, standard	0-0-50	93	96
Sulfate of potash, coarse		72	
Potassium nitrate	13-0-44	78-81	90

[1]The values are average densities which may vary slightly depending on compaction of the product.
[2]Pure salts at 86°F (30°C).

Often it is necessary to determine the percentage of an element in a fertilizer material. Table 5-9 provides easy conversion factors to make the conversion from the compound to the element and from the element to the compound for common fertilizer materials.

SUPPLEMENTARY READING

1. *Commercial Fertilizers.* National Fertilizer Development Center, TVA. 1988.
2. *Fertilizers and Soil Amendments.* R. Follett, et al. Prentice-Hall, Inc. 1981.
3. *Fertilizers and Soil Fertility,* Second Edition. U. S. Jones. Reston Publishing Co. 1982.
4. *Fertilizer Technology and Use,* Third Edition. O. P. Engelstad, ed. Soil Sci. Soc. of America. 1985.
5. *Nitrogen in Agricultural Soils.* F. J. Stevenson, ed. American Society of Agronomy, 1982.
6. *Potassium in Agriculture.* American Society of Agronomy. 1986.

7. *The Role of Phosphorus in Agriculture.* F. E. Khasawneh, E. C. Sample and E. J. Kamprath, eds. American Society of Agronomy. 1980.
8. *Soil Fertility and Fertilizers*, Fourth Edition. S. L. Tisdale, W. L. Nelson and J. D. Beaton. The Macmillan Company. 1985.
9. *Using Commercial Fertilizers*, Fourth Edition. M. H. McVickar and W. M. Walker. The Interstate Printers & Publishers, Inc. 1978.

CHAPTER 6

Fertilizer Formulation, Storage and Handling

With the rapid technological developments of recent years in the fertilizer trade, the user as well as the fertilizer supplier has a wide choice of fertilizer systems available for use. These may be grouped into three broad classifications. First is the system that makes direct application of homogeneous products of uniform size and composition. Such products are available having a wide range of grades and ratios. Second is a system using products from the first category, along with other fertilizer materials in granular form, for use in bulk blending plants to produce ratios and grades readily adaptable to individual needs. Such products have been generally sized to minimize segregation in storage and handling. The third grouping is the fluid fertilizer system that has products ranging from clear liquid solutions to suspensions. It has as its assets ease of handling, uniform composition and adaptability to additions of herbicides and insecticides.

The adoption of these systems has been hastened because of the development of new products and because they provide cost savings at all levels from manufacture to transportation, storage and application. With new technological breakthroughs, substitution or shift from one system to another will occur. This is well illustrated by the development of superphosphoric acid, which has stimulated the expansion of the fluid fertilizer business.

Fertilizer suppliers have several options as to the types of production units they operate and the materials they have available. Factors influencing their choices include capital assets, size of distribution area, availability of raw materials, agronomic suitability of certain products and personal preferences.

FORMULATION

Homogeneous Products

The unique characteristic of a homogeneous product is that each

117

granule or pellet has the same analysis. Homogeneous fertilizer products are usually manufactured in large factories and are supplied to the resellers where they may be used in blends or applied directly to the land. Homogeneous plants utilize ammonia, sulfuric and phosphoric acids and other raw materials.

Some common grades are 18-46-0, 16-20-0, 16-16-16, 19-9-0, 11-55-0, 27-12-0, 6-20-20 and 12-12-12. In addition to the three primary nutrients, nitrogen, phosphorus and potassium, it is also possible to include micronutrients as a guaranteed constituent in the product.

Bulk Blends

Bulk blends are physical mixes of two or more fertilizer materials. The bulk blend plant receives fertilizer products from a basic producer, stores them and blends them together as needed in some type of mixing device. Some of the materials more commonly used to make blends are ammonium nitrate, ammonium sulfate, diammonium phosphate, urea and potash materials. The blends may be taken directly to the field and spread, or, in some cases, they may be bagged. One of the persistent problems with blends is segregation or separation of one component or raw material

Fig. 6-1. Bagging fertilizer in a dry blend plant.

from another. Frequently this means separation of the nitrogen from the phosphate or potash. Consequently, when a blend is spread, uneven distribution of nutrients may occur. The principal contributing factor to segregation is the use of materials of different sizes. For example, if fine crystalline ammonium sulfate is blended with granular diammonium phosphate, segregation occurs due to difference in the size of the two materials. Most fertilizers are screened through a $-6+16$ Tyler screen. If one component such as diammonium phosphate is predominantly $-6+8$ and the other, ammonium sulfate, is predominantly $-12+16$, segregation occurs even though both lie within the $-6+16$ range.

Blenders must exercise care in the selection of raw materials. Also, blenders should avoid allowing piles of finished product to cone, either in storage or when loaded into a truck or trailer. In coning, the larger materials run to the outside of a pile, while the smaller materials stay in the center. Coning can be largely overcome with the use of a flexible spout. Particle shape and density contribute little to segregation problems.

Micronutrients can be added to blends, but a granular form having the same size range as other components is recommended. Another method of incorporating micronutrients is by spraying a solution of the elements on the blend during mixing. The amount that can be applied in this manner is limited by the amount of liquid spray that can be introduced into a dry blend before problems develop.

Liquids and Suspensions

Fluid fertilizers may be classified as liquids and suspensions. Liquids include nitrogen solutions, phosphoric acid and liquid mixes. Nitrogen solutions have already been discussed in Chapter 5.

Liquid mixes are produced by neutralizing phosphoric acid with ammonia. If ortho acid is used, the usual product will be an 8-24-0 grade. Where ammonia and the acid come in contact and are mixed, stainless steel is preferred. Considerable heat is released during ammoniation, and the solution will be hot. If potash is desired, it should be added at this time, as the introduction of potash will lower the solution temperature.

If superphosphoric acid (68 to 76 percent P_2O_5, 50 to 75 percent of which is present as polyphosphate) is ammoniated, it is possible to produce a stable 10-34-0 solution. The higher analysis of this solution is a result of the greater solubility of the pyro, tripoly and other more condensed phosphates. As condensed species are rather unstable at high temperatures, it is necessary to cool 10-34-0 during production. At high temperatures, hydrolysis of the condensed phosphate occurs with reversion to the

Table 6-1

Properties of Ammonium Phosphate Liquids

Grade	8-24-0	9-30-0	10-34-0	11-37-0
Acid used in production	Orthophosphoric	Superphosphoric	Superphosphoric	Superphosphoric
Percent N by weight	8	9	10	11
Percent P_2O_5 by weight	24	30	34	37
Density, lbs./gal. @ 60°F	10.5	11.3	11.4	11.7
N content, lbs./gal.	0.84	1.02	1.14	1.29
P_2O_5 content, lbs./gal.	2.52	3.39	3.87	4.33
Polyphosphates, % of total P_2O_5	none	40 – 45	>50	>65
Viscosity, CP @ 75°F	—	—	73	80
Safe storage temperature, °F	12	0	below 0	0
pH	6.4 – 6.6	6.2 – 6.6	5.8 – 6.1	5.8 – 6.2

ortho form as illustrated in Figure 6-2. When this occurs, the "salting out" temperature is raised, making the solution unstable. Therefore, cooling the solution to below 100°F within an hour after production is recommended.

Low poly (20 percent) superphosphoric acid contains significant quantities of metal ions such as magnesium, iron and aluminum. When the acid is ammoniated in the tee-pipe reactor process, developed at TVA, the heat of reaction between the acid and the ammonia drives off additional combined water and forms a large proportion of condensed phosphate (70 to 80 percent) in the final 10-34-0 product.

Metal salts of long-chain condensed phosphates are more soluble than

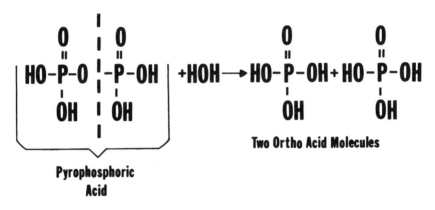

Fig. 6-2. Hydrolysis of pyrophosphoric acid.

ortho and pyro salts. Consequently, they are in solution and do not cause precipitation or sludge problems in the 10-34-0 grade.

Although rather high-analysis N-P grades can be produced as clear liquids, the addition of potash raises the "salt out" temperature, thus making it impossible to produce high-analysis N-P-K grades. There has been a constant increase in the grade of fluids with the progression from orthophosphate to polyphosphate systems.

Suspensions are saturated solutions with crystals of plant nutrients or other materials suspended in the solutions. Usually, a suspending agent such as an attapulgite-type gelling clay is used to suspend the undissolved salt crystals throughout the liquid medium. Suspensions represent a minor portion of the fluid market.

Fig. 6-3. Liquid fertilizer facility.

STORAGE AND HANDLING

To bring efficiency to fertilizer production and distribution, proper and safe storage must be an integral part of the system. Fertilizer plants should operate throughout the year to be efficient. Storage by the producer is necessary to keep the facilities operating, and storage by the supplier is necessary to ensure having a supply available when needed. Fertilizer storage will not be elaborated on here, but those who contemplate storage should make a thorough study of their needs and the facilities available.

Dry Materials

Ammonium nitrate is an excellent fertilizer material that presents no hazard where good storage and proper handling procedures are observed. Precautions to be taken are:

1. Keep it away from open flame.
2. Avoid contaminating it with foreign matter.
3. In case of fire, flood the area with water.

4. Burn empty bags out of doors.
5. Sweep up and dispose of all contaminated material.
6. Do not store in close proximity to steam pipes or radiators.
7. Keep it separate from other materials stored in the same warehouse, especially combustible materials and urea.

Dry urea is totally incompatible with dry ammonium nitrate. Some ammonium phosphates contain small amounts of ammonium nitrate, and the mixing of these materials with urea should be avoided. Ammonium nitrate has a critical relative humidity of 59.4 at 86°F, and under humid conditions it tends to absorb moisture. In areas of high humidity where ammonium nitrate is manufactured, dehumidified storage may be necessary. At small retail outlets and bulk blend plants, a tight bin and a polyethylene cover sheet can be used for storage.

Ammonium nitrate has been used safely for many years. However, if it accidentally becomes contaminated with fuel oil or other combustible materials, it should be disposed of in a safe manner.

Urea — The physical handling and storage of urea is very similar to ammonium nitrate. Urea is nonexplosive and does not have the restrictive regulations imposed on ammonium nitrate. Its critical relative humidity is 72 at 86°F. Thus, it is slightly hygroscopic.

Ammonium sulfate — This material is safe and easy to store. Because of its high critical relative humidity of 81 at 86°F, storage problems are infrequent. The product is corrosive, so concrete or wooden storage structures are preferred.

Phosphates and potash — Except under extremely adverse conditions the ammonium phosphates, straight phosphates and potash materials require no specialized storage. Like most fertilizers, they tend to be corrosive, so concrete and wood are preferred materials for storage structures. Due to the density, large piles or bags of these materials stacked excessively high may cause "setting up," but the lumps are easily broken.

Liquid Materials

Anhydrous and aqua ammonia — These materials are widely used in production agriculture but are inappropriate for most horticulture uses due to handling and safety requirements.

Urea – ammonium nitrate solutions — Two commonly used nonpressure solutions made from urea, ammonium nitrate and water are standardized at 32 percent nitrogen and 28 percent nitrogen. The latter is used during cold weather, since it has a lower salting out temperature. These solutions are used for direct application and for making multinutrient liquid fertilizers by combining them with neutral phosphate solutions and potash.

Ammonium nitrate solution — The common non-pressure solution of ammonium nitrate in water is usually standardized at 20 percent nitrogen content. It is used for direct application or for making multinutrient liquid fertilizers. Some solutions containing higher concentrations of ammonium nitrate are used for manufacturing purposes only, since they need to be kept hot to prevent salting out.

Urea solution — The common urea solution in water contains 23% nitrogen. It is used for direct application or for making multinutrient liquid fertilizers. Some solutions containing higher concentrations are used for manufacturing purposes where, as with ammonium nitrate solutions, they must be kept hot to prevent salting out.

Phosphoric and superphosphoric acids — These acids are widely used by the fertilizer manufacturer and formulator. Both are corrosive, although superacid is somewhat less corrosive than orthophosphoric acid. Rubber-lined or 316 stainless steel is necessary for ortho acid and is recommended for superacid.

Superphosphoric acid is hygroscopic and, if allowed to absorb moisture, will form a thin layer of corrosive ortho acid on the surface. This should be prevented by having a silica-gel breather installed to prevent moisture from entering the tank.

Precautions must be taken against phosphoric or superacid coming in contact with the skin or eyes. These acids are not extremely hazardous, and prompt washing with copious quantities of water is an effective remedy. Superacids of 76 to 83 percent P_2O_5 are strong dehydrating agents; therefore, they have a greater tendency to cause blistering than the less concentrated acids.

Clear liquid and fluid suspensions — Clear liquid fertilizers are easy to handle. If they are neutral solutions, mild steel storage can be utilized and any conventional pumping system can be used.

Suspensions store best in vertical mild steel or plastic tanks equipped for air sparging. When the suspensions are not of good quality, solids may tend to collect on the bottom outside walls of a flat-bottom tank where they cannot be resuspended by air sparging.

Sulfuric acid — This acid is used extensively in the fertilizer industry. It is mostly used by basic producers in the manufacture of phosphatic compounds and ammonium sulfate. Occasionally, it may be used by a dealer or farmer to deal with a specific soil problem.

Sulfuric acid has inherent hazards which limit its use for horticultural purposes. However, there are formulated products that contain dilute concentrations for use in low-volume irrigation systems.

Sulfur materials for formulation of liquids — These materials include ammonium bisulfite and ammonium thiosulfate. For more details on these materials, see Chapter 5.

A major problem of liquid fertilizers is corrosion, and this demands familiarity with the characteristics of various solutions and the metallic composition of storage containers. Storage vessels, pipelines, valves and fittings of all kinds should not contain materials that are destroyed rapidly when they come in contact with the solution.

Fig. 6-4. A modern fertilizer manufacturing unit producing dry and liquid fertilizers.

SUPPLEMENTARY READING

1. *Agricultural Ammonia Safety.* The Fertilizer Institute. 1974.
2. *Fertilizer Nitrogen, Its Chemistry and Technology,* Second Edition. V. Sauchelli. Reinhold Publishing Corp. 1968.
3. *Fertilizers and Soil Fertility,* Second Edition. U. S. Jones. Reston Publishing Co. 1982.
4. *Fertilizer Technology and Use,* Third Edition. O. P. Engelstad, ed. Soil Sci. Soc. of America. 1986.

5. *Nitrogen in Agricultural Soils.* F. J. Stevenson, ed. American Society of Agronomy. 1982.
6. *Physical Properties of Fertilizers and Methods for Measuring Them.* G. Hoffmeister. National Fertilizer Development Center. 1979.
7. *Potassium in Agriculture.* American Society of Agronomy. 1985.
8. *The Role of Phosphorus in Agriculture.* F. E. Khasawneh, E. C. Sample and E. J. Kamprath, eds. American Society of Agronomy. 1980.
9. *Soil Fertility and Fertilizers,* Fourth Edition. S. L. Tisdale, W. L. Nelson and J. D. Beaton. The Macmillan Company. 1985.
10. *Storage of Ammonium Nitrate.* NFPA No. 490. National Fire Protection Association. 1974.

CHAPTER 7

Correcting Problem Soils with Amendments

The soils under consideration in this chapter are problem soils that require special management and remedial measures. The primary concern is with the use of amendments such as lime, gypsum, sulfur and other materials which, when properly used, make the soil more productive.

Those soils which have excess acidity may produce toxicity in plants from solubilized aluminum and manganese; alter the populations and activities of the microorganisms involved in nitrogen, sulfur and phosphorus transformations in the soil; and thus affect the availability of nutrients to higher plants. Acidity may affect the growth of plants directly. Usually it is an indication of a low level of calcium and magnesium. The use of lime as an amendment can ameliorate many or all of these problems.

Landscape uses often include planting of deep-rooted shrubs and trees. Shallow fills of good topsoil are often underlain by old surface and subsoil which have vastly different texture and pH. In such cases it is important to sample the subsoil down to the effective rooting depth to check for incompatible conditions, such as acidity, alkalinity or excess salt.

Those soils which have a high pH, often referred to as *alkaline soils,* may be afflicted with high sodium content, excess salts, poor structure and other problems. The treatment of these soils with amendments requires a completely different approach. This chapter will be concerned with the identification of the problems and the appropriate measures to take in their correction.

ACID SOILS

Acid soils are generally found in areas of heavy rainfall, on sandy or light textured soils, or where high rates of acid-forming fertilizers have been used on poorly buffered soils and where peat or organic soil deposits occur. Landscape cuts often expose poorly buffered subsoils that may be prone to acidification when reclaimed for residential and other uses (golf courses, parks, etc.). Acid soils are more readily formed from weathering of acid igneous rocks, such as granites, and secondary rocks, such as sandstones, than from basic igneous rocks, such as basalts.

Soils become acid because the cations on the soil colloids, primarily calcium, are replaced by hydrogen ions. This implicates ion exchange and other adsorptive reactions that are associated with the colloids, such as the type of clay and the amount of organic matter. Aluminum also comes into play to the extent that highly acid soils can result from aluminum saturation.

The nature of soil acidity is complex; one part, called the *active acidity*, is made up of the hydrogen ions in the soil solution. These are the ions measured when the pH of a soil is determined. Another and much larger part of the total acidity is usually referred to as the *potential acidity*, representing the hydrogen ions held on the colloidal surface. Since clays and organic particles have large surface areas, they usually have a much higher total acidity than sandy soils, and it takes more lime to change the pH of these soils.

Table 7-1

Plants Grouped According to Their Tolerance to Acidity

Very Sensitive to Acidity	Will Tolerate Slight Acidity	Will Tolerate Moderate Acidity	Strong Acidity Favorable
Sweet clover	Red clover	Vetch	Blueberry
Cabbage	Mammoth clover	Rye	Cranberry
Cauliflower	Alsike clover	Buckwheat	Holly
Lettuce	White clover	Millet	Rhododendron
Onion	Timothy	Sudan grass	Azalea
Spinach	Kentucky bluegrass	Redtop	
Asparagus	Pea	Bentgrass	
Beet	Carrot	Potato	
Parsnip	Cucumber	Parsley	
Celery	Brussels sprouts	Sweet potato	
Muskmelon	Kale		
	Kohlrabi		
	Pumpkin		
	Radish		
	Squash		
	Lima, pole and snap beans		
	Sweet corn		
	Tomato		
	Turnip		

Table 7-2

Preferred pH Ranges of Various Plantings[1]

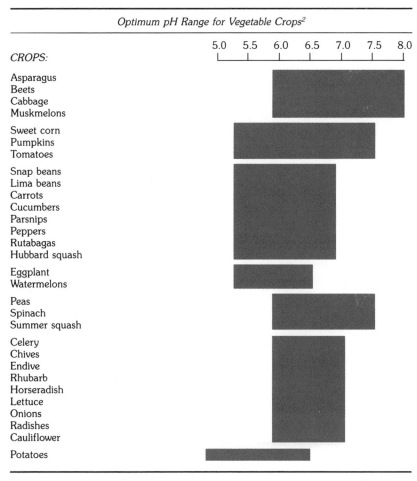

Optimum pH Range for Vegetable Crops[2]

CROPS:	5.0	5.5	6.0	6.5	7.0	7.5	8.0
Asparagus Beets Cabbage Muskmelons							
Sweet corn Pumpkins Tomatoes							
Snap beans Lima beans Carrots Cucumbers Parsnips Peppers Rutabagas Hubbard squash							
Eggplant Watermelons							
Peas Spinach Summer squash							
Celery Chives Endive Rhubarb Horseradish Lettuce Onions Radishes Cauliflower							
Potatoes							

(Continued)

Table 7-2 (Continued)

	Optimum pH Range for Fruit and Nut Crops								
CROPS:	4.0	4.5	5.0	5.5	6.0	6.5	7.0	7.5	8.0
Blueberries									
Strawberries									
Blackberries[2] Raspberries[2]									
Apples Apricots Cherries Grapes Peaches Plums Pecans									
Walnuts									
Currants Gooseberries									

[1] Source: *Fertilizing Gardens in Kansas.* Coop. Ext. Ser. Kansas State Univ. July 1975.

[2] Information reprinted from *Producing Vegetable Crops,* Third Edition. G. W. Ware and J. P. McCollum. The Interstate Printers & Publishers, Inc. 1980.

The pH measurement is commonly made when testing soils. Although it is a useful index, it is often misunderstood and misused. The measurement is usually made on a 1:1 mixture of soil and water or on a saturated soil paste. Differences of several tenths of a pH unit higher are observed as the water:soil ratio is increased. The interpretation of the pH value must be based upon more than a simple reading.

The pH range for acid soils ordinarily is from 4.0 to 7.0. Values much below 4.0 are obtained only when free acids, such as sulfuric acid, are present. Values above 7.0 indicate alkalinity.

Plant tolerance to soil acidity varies widely. In some cases, plants even prefer strong acidity. See Tables 7-1 and 7-2.

Lime Requirement

Different methods have been developed to determine the amount of lime needed to bring the pH of an acid soil to a desirable range. All of those presently used take into consideration the soil texture and organic matter content and use a specialized procedure. Table 7-3 shows what the effect of finely ground limestone is on different soils.

Table 7-3

Approximate Amount of Finely Ground Limestone Needed to Raise the pH of a 7-Inch Layer of Soil[1]

| | Lime Requirements, Lbs. / 1,000 Sq. Ft. | |
Soil Texture	From pH 4.5 to 5.5	From pH 5.5 to 6.5
Sand and loamy sand	23	28
Sandy loam	37	60
Loam	55	78
Silt loam	69	92
Clay loam	87	106
Muck	174	197

[1] Adapted from USDA Agricultural Handbook No. 18.

Liming Materials

The commonly used materials for liming soils are the carbonates, oxides, hydroxides and silicates of calcium and magnesium. A material does not qualify as a liming compound just because it contains calcium and magnesium. Gypsum, for example, has little direct effect on soil pH and cannot be used to correct a low soil pH. Table 7-4 gives the liming materials most commonly used to treat acid soils.

The value of a liming material is governed by molecular composition, purity and degree of fineness. Some established standards indicate that all material must pass through a 60-mesh screen to have a full efficiency rating.

Table 7-4

Common Liming Materials[1]

Name	Chemical Formula	Equivalent % CaCO$_3$	Source
Shell meal	$CaCO_3$	95	Natural shell deposits
Limestone	$CaCO_3$	100	Pure form, finely ground
Hydrated lime	$Ca(OH)_2$	120 – 135	Steam burned
Burned lime	CaO	150 – 175	Kiln burned
Dolomite	$CaCO_3 \cdot MgCO_3$	110	Natural deposit
Sugar beet lime[2]	$CaCO_3$	80 – 90	Sugar beet by-product lime
Calcium silicate	$CaSiO_3$	60 – 80	Slag

[1] Ground to same degree of fineness and at similar moisture content.

[2] From 4% to 10% organic matter.

The benefits of liming acid soils are broad in scope. Calcium and magnesium are essential plant nutrients, and their addition may provide direct value. Additionally, the correction of chemical, physical and biological conditions may result in striking improvements in plant growth. Phosphorus availability is greatest at a soil pH of 6.5 to 7.0. Toxicities of elements are minimized at pH values of 6.0 to 7.0, and availability of the micronutrients is optimized. Biological activity is favored at a neutral or near neutral pH, including processes such as nitrification, nitrogen fixation and decomposition of plant residues. At the same time, soil aggregation and good structural development are favored.

SALINE AND ALKALI SOILS

Saline soils generally occur in regions with arid or semiarid climates. In humid areas, rainfall is usually sufficient to move the soluble salts out of the soil into drainage waters, making the incidence of salinity rare. However, the intrusion of sea or brackish water may induce soil salinization, especially in low-lying areas. Arid regions frequently are inadequately drained and are subject to high evaporation rates, thus allowing salt

Fig. 7-1. Salt buildup in a California vineyard.

buildup to occur. Often irrigation is practiced where there are no drainage outlets, and this results in salinization and alkalization.

The soluble salts that occur come indirectly from the weathering of primary minerals and from waters which carry salts from other locations. For example, the Colorado River at Yuma, Arizona, carries more than 1 ton of salt per acre-foot of water. Using this water will result in rapid buildup of salt unless adequate drainage is provided and proper irrigation practices are used. Many ornamental plant species are particularly susceptible to both salinity and alkalinity. Symptoms include leaf burn, dieback and a generally unhealthy appearance, assuming the plants survive at all.

Alkali soils contain excessive amounts of sodium. When soils come in contact with waters containing a high proportion of sodium, this cation becomes dominant in the soil solution and replaces calcium and magnesium on the clay. As a consequence of this adsorption of sodium, alkali soils are formed. The use of softened or gray water (treated waste effluent) for irrigation can cause a rapid buildup of sodium. As a result, soil structure and water penetration will suffer.

SALINE SOILS

The term *saline* is applied to soils which have a conductivity of the saturation extract greater than 4.0 dS / m, and the exchangeable sodium percentage is less than 15 percent. These soils normally have a pH value below 8.5 and have good physical properties. Saline-alkali soils sometimes contain gypsum. If this is the case, leaching will help dissolve calcium, and the soil will provide its own gypsum amendment. Free calcium carbonate (limestone) may occur in these soils. The selection of amendments in this case can also be made from those that solubilize calcium carbonate and form calcium sulfate right in the soil. Such amendments are soil sulfur, sulfuric acid, ferric and ferrous sulfate and aluminum sulfate. Organic materials may also produce acids during decomposition and, therefore, provide soluble calcium to improve soil structure to aid water penetration and salt leaching.

NONSALINE-ALKALI SOILS (SODIC SOILS)

The term *nonsaline-alkali* or *sodic* is applied to soils which have conductivity of the saturation extract below 4.0 dS / m, and the exchangeable sodium percentage is greater than 15 percent. They normally have a pH value greater than 8.5 and are characterized particularly by their poor physical structure. Sodic soils contain sufficient exchangeable sodium to interfere with the growth of most plants. These soils are commonly termed *alkali, black alkali* and *slick spot* soils. The darkened appearance is caused by the dispersed and dissolved organic matter deposited on the soil surface by evaporation.

Adsorbed sodium causes disintegration of the soil aggregates, dispersing the soil particles and effectively reducing the large pore space. This makes leaching difficult, since the soil becomes almost impervious to water.

On non-calcareous soils, treatment must be made with gypsum or other soluble calcium salts. On calcareous soils, treatment may be with gypsum or acidifying materials. Calcium replaces sodium on the clay surface and helps bring about a better physical condition that will allow sodium and excess salts to be leached. Organic materials such as manure and crop residues may be helpful by providing a better physical condition for leaching.

Reclamation of alkali land must be measured in terms of the time and cost of the treatment, the intended use and the ultimate value of the land. The necessary treatment for deep-rooted plants may vary from that for turf or other shallow-rooted ground cover. Because of the special soil conditions which may exist, it is advisable to consult with soil specialists for specific recommendations to fit the individual case.

SOIL AMENDMENTS

Identification of the specific problems is necessary before the amendment is chosen. Also, physical soil problems such as compaction layers (hardpan, plowpan, etc.) which are not directly affected by the amendment must be corrected before the amendment can be effective. Often efforts are made to amend a saline or alkali soil without awareness that the problem may be compounded by high boron content in the soil or the leaching waters. Also, there are those who claim that some special material or process will "magically" cure the problem. Rarely does low-rate – one-time application of even the most suitable material provide any significant lasting improvement. The principles concerning the use and selection of soil amendments are well known, and no shortcut to the proper application of these principles will bring any effective or lasting benefits.

Soil Considerations

The presence of lime (free calcium carbonate) in the soil allows the widest selection of amendments. To test for this, a simple procedure can be followed by taking a spoonful or clod of soil and dropping a few drops of muriatic or sulfuric acid on it. If bubbling or fizzing occurs, this indicates the presence of carbonates or bicarbonates.

If the soil contains lime, any of the amendments listed in Table 7-5 may be used. If lime is absent, select only those amendments containing soluble calcium.

Table 7-5

Commonly Used Materials and Their Equivalent Amendment Values

Material (100% Basis)	Chemical Formula	Pounds of Amendment Equivalent to	
		100 Pounds of Pure Gypsum	100 Pounds of Soil Sulfur
Gypsum	$CaSO_4 \cdot 2H_2O$	100	538
Soil sulfur	S	19	100
Sulfuric acid (conc.)	H_2SO_4	61	320
Ferric sulfate	$Fe_2(SO_4)_3 \cdot 9H_2O$	109	585
Lime sulfur (22% S)	CaS_x	68	365
Calcium chloride	$CaCl_2 \cdot H_2O$	86	—
Calcium nitrate	$Ca(NO_3)_2 \cdot H_2O$	106	—
Aluminum sulfate	$Al_2(SO_4)_3 \cdot 18H_2O$	129	694
Ammonium polysulfide	$(NH_4)_2S_x$	4	23

The percent purity is given on the bag or identification tag.

Types of Amendments

Calcium-containing amendments, such as gypsum, react in the soil as follows:

gypsum + sodic soil   ⟶   calcium soil + sodium sulfate

Leaching is essential in removing the sodium salt, the amount dependent upon the severity of the alkali problem.

Acids such as sulfuric acid require two steps:

1. sulfuric acid + lime   ⟶   gypsum + carbon dioxide + water
2. gypsum + sodic soil   ⟶   calcium soil + sodium sulfate

The acid-forming materials such as sulfur go through three steps. First is oxidation to form the acid, then steps 2 and 3 are the same as 1 and 2 above.

Table 7-6

Approximate Amounts of Soil Sulfur (95%) Needed to Increase the Acidity of the 6-Inch-Deep Layer of a Carbonate-Free Soil

Change in pH Desired	Pounds of Sulfur per 1,000 Sq. Ft.		
	Sand	Loam	Clay
8.5 to 6.5	45.9	57.4	68.8
8.0 to 6.5	27.5	34.4	45.9
7.5 to 6.5	11.5	18.4	23.0
7.0 to 6.5	2.3	3.4	6.9

1. sulfur + oxygen + water $\longrightarrow$ sulfuric acid
2. sulfuric acid + lime $\longrightarrow$ gypsum + carbon dioxide + water
3. gypsum + sodic soil $\longrightarrow$ calcium soil + sodium sulfate

Approximate rates of sulfur for varying soil conditions are given in Table 7-6.

Effectiveness of Amendments

The values given in Table 7-5 are for 100 percent pure amendments. If an amendment is not pure, a simple calculation will indicate the amount needed to be equivalent to pure material:

$$\frac{100}{\% \text{ purity}} = \text{amount of pure material}$$

Example: If gypsum is 80 percent pure, the calculation would be 100 / 80 = 125 lbs., or 125 lbs. of 80 percent gypsum would be equivalent to 100 lbs. of 100 percent pure gypsum.

When considering sulfur, the purity and degree of fineness must be taken into account. Most sulfur is over 95 percent pure. Since sulfur must be oxidized before it is effective as an amendment, the finer the material, the faster it will be oxidized in the soil, since greater surface area is exposed. Finely divided sulfur particles will oxidize within one season, while coarse materials may take years to become oxidized.

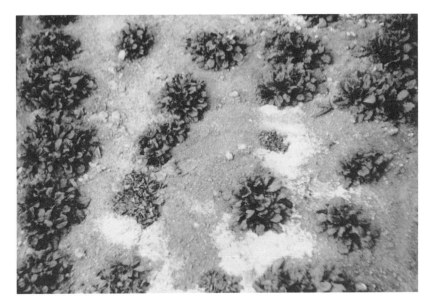

Fig. 7-2. Ajuga growth severely affected by salt.

MANAGEMENT OF SALINE
AND SODIC SOILS

Often it is not practical to completely reclaim saline or sodic soils or even to maintain these soils at a low saline or sodic condition. The reasons may be cost of reclamation, inability to adequately drain, high cost of amendments, low-quality irrigation water, etc.

Management practices that aid in the control of salinity and alkalinity include:

1. Selection of crops or crop varieties that have tolerance to salt or sodium.
2. Use of special planting procedures that minimize salt accumulation around the seed, shrubs or trees.
3. Use of sloping beds or special land preparation procedures and tillage methods that provide a low-salt environment for the germinating seed.
4. Use of irrigation water to maintain a high water content to dilute the salts or to leach the salts out of the germination and root growing zone.

5. Use of physical amendments for improving soil structure. (See Chapter 9.)
6. Deep fracturing of the soil to break up hardpan or other impervious layers to provide internal drainage.
7. Use of chemical amendments as described.
8. Establishment of proper surface and internal drainage.

It is essential to know the nature of soil, both physical and chemical, the quality and quantity of irrigation water available, the climate of the area including the growing season, the economics of the situation, etc., before a satisfactory management program can be developed. Consulting with the appropriate authorities and having suitable tests made are essential steps in reclamation and management.

SUPPLEMENTARY READING

1. *Chemistry of Irrigated Soils.* R. Levy, ed. Van Nostrand Reinhold Company. 1984.
2. *Diagnosing Soil Salinity.* USDA Agricultural Information Bulletin 279. 1963.
3. *Diagnosis and Improvement of Saline and Alkali Soils.* USDA Agricultural Handbook No. 60. 1954.
4. *Drip / Trickle Irrigation in Action.* Vols. I and II. Proceedings of the Third International Drip / Trickle Irrigation Congress. American Soc. of Agric. Engineers. 1985.
5. *Gypsum and Other Chemical Amendments for Soil Improvement.* University of California, Extension Leaflet 149. 1962.
6. *Irrigation Principles and Practices,* Fourth Edition. V. E. Hansen and G. E. Stringham. John Wiley & Sons, Inc. 1980.
7. *Reclaiming Saline and Alkali Soils.* Extension Publication, Fresno. 1972.
8. *Saline and Sodic Soils.* E. Bresler, B. L. McNeal and D. L. Carter. Springer-Verlag. 1982.
9. *Soil, Yearbook of Agriculture.* U.S. Government Printing Office. 1957.
10. *Soil Acidity and Liming.* Agronomy Monograph 12. American Society of Agronomy. 1967.
11. *Soil Reclamation Processes.* Tate and Klein. Marcel Dekker, Inc. 1985.
12. *Soil Salinity Under Irrigation.* I. Shainberg and J. Shalhevet, eds. Springer-Verlag. 1984.
13. *Soil Survey Manual.* USDA Agricultural Handbook No. 18. Soil Survey Staff. U.S. Government Printing Office. 1951.
14. *Soil Testing and Plant Analysis.* L. M. Walsh and J. D. Beaton, eds. Soil Sci. Soc. of America. 1973.

CHAPTER 8

Soil Organic Matter

Organic matter is one of the major keys to soil productivity. Western soils differ considerably in organic matter content from region to region. Since western climate is mostly semiarid, the average organic matter content of field soils is quite low, usually less than 1 percent. Soil organic matter content in the higher rainfall areas, having an abundance of vegetation, may range up to 8 to 10 percent. River delta soils containing centuries-old accumulations of aquatic vegetation, reeds and sedges may be predominantly organic soils (peat and muck). Artificially amended growing media contain organic matter ranging from below 10% to above 80%.

Soil organic matter consists of plant and animal residues in various stages of decomposition, living soil organisms and substances synthesized by these organisms. The amount of organic matter that may accumulate in a soil from plant tissues depends upon the temperature, moisture, aeration, soil pH, microbial population and the quantity and chemical nature of the plant residue returned to the soil (Figure 8-1).

The chemical composition of soil organic matter is categorized in three major groups: *polysaccharides, lignins* and *proteins* (Table 8-1). Besides these three groups, a variety of other substances such as fats and waxes also occur in plant residues. The polysaccharides include cellulose, hemicelluloses, sugars, starches and pectic substances. Lignins are complex materials derived from woody tissues of plants. Proteins, the principal nitrogen-containing constituents of organic matter, exist in all life forms. These three classes of materials are sources of food for soil micro-organisms.

Organic residue is decomposed in the soil by living organisms, primarily bacteria, fungi, and actinomycetes. Each group of organisms becomes a dominant factor at various stages of decomposition. These and larger organisms such as earthworms and insects ingest organic residue and soil, thereby binding together soil particles into stable aggregates. Steam sterilization of growing media slows or totally stops decomposition of organic materials. Practically every soil property is affected by soil organic matter.

The dark-colored organic residue which resists further decomposition

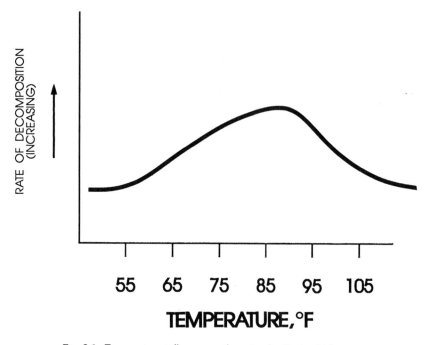

Fig. 8-1. Temperature influence on the rate of soil microbial activity.

is referred to as *humus*. Humus increases friability of soils, improves tilth and facilitates aeration and water penetration.

PRINCIPAL BENEFITS OF
SOIL ORGANIC MATTER

Soil organic matter:

1. Helps build stable soil aggregates, thus improving soil structure and tilth.
2. Improves aeration and water penetration.
3. Improves moisture-holding capacity.
4. Provides an abundance of negatively charged colloidal particles (humus) capable of holding and exchanging nutritive cations.
5. Acts as a buffering agent by decreasing the tendency for an abrupt pH change in the soil when acid- or alkaline-forming substances are added.

Table 8-1

Common Constituents of Soil Organic Matter and Relative Rates of Decomposition in Non-sterilized Soil

Organic Constituents	Approximate Percent of Total Organic Matter	
		Rapidly decomposed
Sugars, starches, simple proteins	1 – 5	
Crude proteins	5 – 20	
Hemicelluloses	10 – 25	
Cellulose	30 – 50	
Lignins, fats, waxes	10 – 30	Very slowly decomposed

6. Affects the formation of metal-organic complexes, thus stabilizing soil micronutrients that otherwise might not be available.
7. Provides a source of many plant nutrients.

It is important to recognize that the major plant nutrients may not exist in soil organic matter in sufficient quantity to sustain maximum plant growth. Soil organic matter usually contains 5 or 6 percent nitrogen and lesser amounts of phosphorus. These nutrients must be mineralized to the inorganic form during decomposition before they become available. There is some evidence that small quantities of organic compounds can be absorbed directly by plants without mineralization. The contribution is small, however, in the overall nutrition of plants. A simplified equation showing the mineralization process is:

$$\text{organic matter} + H_2O + O_2 \xrightarrow{\text{(soil organisms)}} CO_2 + H_2O + NH_3 + H_3PO_4 + H_2S + \text{other gases} + \text{energy}$$

Organic matter is the source of energy for soil organisms as they multiply and carry on their life processes. Ammonia produced during mineralization may be nitrified to NO_3 (nitrate) by nitrifying bacteria.

Soil microbes require nutrient elements just as do plants. In the process of breaking down an abundant supply of organic matter, a rapidly growing population of microbes will rob the soil of available nitrogen. This may temporarily reduce plant growth if the soil supply of nitrogen is not sufficient to take care of the needs of both the microbes and growing plants. This situation often occurs when a planting soil has been amended with low-nitrogen organic materials such as sawdust or bark fines. By adding nitrogen fertilizer, both the growing plant and soil organisms can have a plentiful supply to meet their needs.

THE RECYCLING OF PLANT NUTRIENTS

The recycling of nitrogen from organic residues to soil to growing plants is a part of the soil nitrogen cycle. (See Chapter 4.)

The phosphorus cycle is similar in some respects to the nitrogen cycle. The effect of soil organic matter on providing available phosphorus to growing plants is also significant. For example, when the carbon:phosphorus ratio is wide, immobilization of available phosphorus occurs. When the ratio is narrow, net increases of available inorganic phosphorus are produced.

Humus adsorbs phosphate ions. These ions are, however, more available to growing plants than phosphorus precipitated as insoluble compounds. Thus, soils high in humus usually contain greater quantities of available phosphorus than soils low in humus.

SOURCES OF ORGANIC MATERIALS

Organic matter in most soils is derived from plant residues and manures. Residues which contain the lowest amount of carbon in relation to nitrogen come from cover crops (green manure): legumes, grasses and mustards. These crops decompose rapidly and provide nutrients in excess of microbial needs. Animal manures can also be valuable sources of humus, but, since they contain salts, precaution must be taken when they are used to avoid excessive salt accumulation.

Residues from grain straw and cotton motes are low in nitrogen but high in carbon and thus take longer to decompose. The addition of nitrogen fertilizer to these residues speeds up decomposition and helps to satisfy the microbial demand for the nutrients. When these residues are incorporated, a rule of thumb is to apply 20 pounds of nitrogen per ton of residue.

Currently, organic matter in the form of animal manures and sewage sludge is used as a nitrogen source and soil amendment (Tables 8-2 and 8-3). If organic wastes completely replaced inorganic nitrogen sources, sol-

Table 8-2

Composition of Manures and Waste Materials[1]

Source	Percent Moisture	N		P₂O₅		K₂O	
		%	Lbs. / Ton	%	Lbs. / Ton	%	Lbs. / Ton
Beef feedlot	68	0.71	14.2	0.64	12.8	0.89	17.8
Dairy	79	0.56	11.2	0.23	4.6	0.60	12.0
Liquid dairy	91	0.24	4.8	0.05	0.1	0.23	4.6
Swine	75	0.50	10.0	0.32	6.4	0.46	9.2
Liquid swine	97	0.09	0.2	0.06	0.1	0.08	0.2
Horse	70	0.69	13.8	0.23	4.6	0.72	14.4
Sheep	65	1.40	28.0	0.48	9.6	1.20	24.0
Poultry (no litter)	54	1.56	31.2	0.92	18.4	0.42	8.4
Liquid poultry	92	0.16	3.2	0.04	0.8	0.29	5.8

[1] Adapted from L. S. Murphy in *Fertilizer Solutions* magazine. March-April 1972. See Table 8-3 for a more extensive list.

uble salts would accumulate when higher quantities of manure were applied the first few years. Corrective measures would have to be taken, and in some moderately saline soils, yields would probably be reduced.

Where sawdust and other similar low-nitrogen mulching materials are incorporated into greenhouse, field or container growing media, chemical nitrogen should be added to bring the total nitrogen level to 1.75 percent by weight of the organic material used. It may not be necessary to add extra nitrogen to composts or organic materials with nitrogen contents above this level. Required nitrogen may also be applied during the growing period via top dressing dry nitrogen fertilizer and / or via irrigation water (sprinkler, spray, drip tube or subirrigation).

ORGANIC CONCENTRATES

Activated sewage sludge, tankage, dried blood, and fish and seed meals are referred to as *organic concentrates* and are used as specialty feed supplements. Untreated plant residues, however, contain less nitrogen and are used as mulches or organic amendments. They contain a higher percentage of nitrogen than most crop residues. Most of the nitrogen is water-insoluble and is made available by microorganisms during decomposition. Synthetic organic fertilizers such as ureaforms and methylene ureas contain significantly higher percentages of nitrogen than natural organic con-

Table 8-3

Average Nutrient Analysis of Some Organic Materials

	N	P_2O_5	K_2O
	-------------------------	(%)	-------------------------
Fresh manure with normal quantity of bedding or litter			
Duck	1.1	1.45	0.50
Goose	1.1	0.55	0.50
Turkey	1.3	0.70	0.50
Rabbit	2.0	1.33	1.20
Bulky organic materials			
Alfalfa hay	2.5	0.50	2.10
Bean straw	1.2	0.25	1.25
Grain straw	0.6	0.20	1.10
Cotton gin trash	0.7	0.18	1.19
Seaweed (kelp)	0.2	0.10	0.60
Winery pomace (dried)	1.5	1.50	0.75
Organic concentrates			
Dried blood	12.0	1.5	—
Fish meal	10.4	5.9	—
Digested sewage sludge	2.0	3.0	—
Activated sewage sludge	6.5	3.4	0.3
Tankage	7.0	8.6	1.5
Cottonseed meal	6.5	3.0	1.5
Bat guano	13.0	5.0	2.0
Bone meal[1]	<1.0	12 – 14	—

[1] Bone meal values vary widely because of moisture content and processing. Available P_2O_5, 12% – 14%; insoluble P_2O_5, 14% – 16%; total P_2O_5, 26% – 28%.

centrates. The process of nitrogen release and availability from synthetic organic fertilizers is the same as the process with natural organics (see Chapter 5).

ORGANIC SOIL AMENDMENTS

Those organic amendments most commonly used by home gardeners and horticulturists are digested sawdust, wood shavings, ground bark, leaf molds, peats and composts. Wood residues can be incorporated directly or composted. Because they have a high moisture-holding capacity, they make excellent soil mulches for blueberries, strawberries, fruit trees, ornamentals and garden crops. Since woody tissue is low in plant nutrients, extra nitrogen and phosphate are needed when they are composted or added to soil.

Peats may be divided into two main types, according to the kinds of plants from which they were derived. Sphagnum peat (peat moss) is derived from species of the sphagnum plant, the remains of which have accumulated over centuries in bogs in parts of Canada and the northern United States. Hypnum peat is derived under similar anaerobic conditions from sedges, reeds, mosses or trees.

Sphagnum peats have a low ash content (below 5 percent), are very low in plant nutrients and are very acid (pH values between 3.0 and 4.5). They do have a very high water-holding capacity equal to 15 to 30 times their own weight, but this is cut in half following drying. They are beneficial as mulches for acid-loving plants; otherwise they may need to be neutralized with limestone (Table 8-4).

Hypnum peats are composed of more lignin-like substances quite resistant to further decomposition. Their dry matter contains 5 to 40 percent ash. They do not exhibit as high a water-holding capacity as sphagnum peat moss and are less acid.

MAKING COMPOST

Composting has become a popular method of utilizing organic wastes for conversion to beneficial organic soil amendments. Composting can be carried out on any scale from large industrial waste management to backyard gardening. Almost any natural organic material can be composted

Table 8-4

Typical Analyses for Organic Soil Amendments

Organic Amendment	Moisture Retention	pH	Organic Matter	Ash
	(% total volume)		(%)	(%)
Sphagnum peat moss	60 – 70	3.2 – 4.5	95 – 99	1 – 5
Hypnum peat moss	55 – 65	4.4 – 6.7	70 – 85	15 – 30
Reed and sedge peat	50 – 60	4.5 – 7.0	85 – 95	5 – 15
Woody peat	30 – 40	3.6 – 5.5	75 – 90	10 – 25
Sawdust	30 – 40	3.8 – 8.0	95 – 99	1 – 5
Ground bark	30 – 40	4.0 – 8.0	90 – 95	5 – 10
Compost	20 – 30	4.0 – 8.0	80 – 85	15 – 20
Leaf mold	20 – 30	4.0 – 7.0	50 – 75	25 – 50

with proper care — cornstalks, straw, leaves, grass clippings, winery refuse and garbage; the microbes are not selective.

Compost heaps should be built no more than 6 feet high so that air can enter at the bottom of the pile. Residue should be cut less than 6 inches in length. Nitrogen fertilizer should be added at a rate of ¼ pound of nitrogen per cubic foot of dry material.

Moisture content between 50 and 70 percent is desirable; decomposition is slowed down when the heap is drier. Anaerobic conditions exist, particularly at the bottom of the heap, where it is wetter. Water and nitrogen are best applied to the layers as the pile is built up. If the moisture becomes excessive it can be reduced by loosening the pile. Allowing temperatures to build up will hasten the decomposition. High-temperature-activated organisms complete the composting process.

In farm and garden practices, the compost should be turned every three or four days. Under favorable conditions, composting can be complete in three or four weeks. The compost is complete when it has cooled, has a dark color and is crumbly and odorless. Composts are particularly beneficial for soils low in organic matter and where frequent tillage and complete removal of crops may lead to soil deterioration.

Composts, like other organic amendments, benefit the soil principally by improving soil structure, water penetration and moisture-holding capacity. They do not contain enough nutrients to satisfy growing plants unless they are fortified with fertilizer prior to mixing with the soil.

NUTRITIONAL VALUE

Food grown "organically" has no more nutritional value than food grown with chemical fertilizers, according to a report of the Institute of Food Technologists. Plants use the nutrients in the soil for their own growth and maturation, and it makes no difference from which source these nutrients come.

Everything the plant uses, and everything thus passed on to those who eat the plant or its fruit, is broken down into molecules; this process takes place equally well whatever the source of the nutrients.

The IFT summary reported that "the term 'organic' has taken on a new connotation referring to food from soil that has been treated only with animal manure and composted materials," in contrast to the dictionary definition which implies that all food is organic because it is the product of living organisms, plant or animal.

The report also discussed the recycling of municipal wastes to solve the problems of waste disposal and nutritional deficiencies. It concluded that it is not feasible to transport large amounts of wastes from cities to plant production areas and that such wastes often contain metals that are poisonous to plants.

According to one authority, Harvard nutritionist Jean Mayer, the so-called "organic" foods may escape pollution by chemicals, but they tend to become the most biologically polluted of all foods.

"Organic fertilizers of animal or human origin are obviously the most likely to contain gastrointestinal parasites," Mayer said.

SUPPLEMENTARY READING

1. *Ball Red Book,* Fourteenth Edition. V. Ball, ed. Reston Publishing Co. 1984.
2. *Chemistry of the Soil.* F. E. Bear. Reinhold Publishing Corp. 1965.
3. *Fundamentals of Horticulture,* Fourth Edition. J. B. Edmond, A. M. Musser and F. S. Andrews. McGraw-Hill Book Co. 1975.
4. *Horticultural and Agricultural Uses of Sawdust and Soil Amendments.* Technical Bulletin. P. Johnson and O. A. Matkin. Copyright 1968 by Paul Johnson.
5. *The Nature and Properties of Soils,* Eighth Edition. N. C. Brady. The Macmillan Company. 1974.
6. *Peat and Muck in Agriculture.* M. S. Anderson, S. F. Blake and A. L. Mehring. Circular No. 888, U.S. Department of Agriculture. 1951.
7. *Soil, Yearbook of Agriculture.* U.S. Government Printing Office. 1957.
8. "Using Organic Wastes as Nitrogen Fertilizers." P. F. Pratt, F. E. Broadbent and J. P. Martin. *California Agriculture,* 27:6, 1973.

CHAPTER 9

Amending the Physical Properties of Soils for Planting and Potting Purposes

For centuries, cover crops, crop residues, animal manures and other organic materials have been applied to the soil to help maintain or improve soil physical properties and productivity. One of the earliest recorded instances of adding manure to the soil was in 900 B.C., as recorded in the Greek epic poem *The Odyssey.*

From that time to the present, the physical condition of soils has been improved by the addition of soil amendments. As we continue to grow and harvest plants, the need for soil improvement will always be present. This is particularly true today as urbanization forces us to use less than ideal soils, and as we concentrate our food and ornamental growing into small areas, greenhouses and containers (Figure 9-2).

The availability and types of amendments to improve the physical properties of soils continually change. Often these materials are by-products. Organic wastes such as bark, sawdust, manure, sludge and mushroom composts are examples. Supplies of rice hulls, cedar and other wood residue are increasing, while redwood and peat moss supplies are declining. Other materials will undoubtedly come into use in the future.

DEFINITION OF PHYSICAL SOIL AMENDMENTS

For the purposes of this chapter, we have defined "physical soil amendments" as any substances used for the purpose of promoting plant growth or improving the quality of crops by conditioning soils solely through physical means. This centers primarily on soil improvement by improving water retention, permeability and air movement.

149

Fig. 9-1. Composting of municipal sludge.

NEED FOR PHYSICAL SOIL AMENDMENTS

The soil is very similar to the foundation of a building. The question to be asked in this case is "Will this soil support good plant growth?" If not, it will need to be amended.

When a soil is adequately supplied with air, water, organic matter and nutrients, little amending is necessary. Such an ideal soil is rare. Generally, it is necessary to add a soil amendment to improve the physical condition of the soil and maintain it for good plant growth.

Replacing the existing soil with a good topsoil would be more costly than amending. Also, unless careful measures are followed, soil layering often results. This can restrict air, water and root movement.

Selecting a proper amendment is very important when planting trees and shrubs. Plants are expected to beautify an area over extended periods of time. However, once an area has been planted it is impractical to amend a soil. This should be done before plants are established.

Fig. 9-2. Many plants are grown in small areas (limited rooting volume). Skillful management is required because small errors can lead to big problems.

Fig. 9-3. Calendulas grown in amended soil on left and unamended soil on right.

SOIL PHYSICAL PROPERTIES

The physical properties of soils are related to soil textures as illustrated in Table 9-1.

Sandy soils have properties that are almost directly opposite those of clay soils. Permeability or porosity is high in a sandy soil and low in a clay soil. While these differences are recognized in maintenance practices, e.g., irrigation, often they are overlooked at the time of amending and planting. For optimum results, it is essential that soil texture be considered when the choice of amendments is made.

Table 9-1

Soil Physical Properties Related to Soil Texture

Soil Texture	Permeability	Water Retention
Sand	High	Low
Loam	Medium	Medium
Silt	Low	High
Clay	Low	High

LONGEVITY OF AMENDMENTS

None of the organic compounds of plant residues are indestructible, although some are very resistant to decomposition. Considering amendments as a whole, their longevity in the soil depends primarily upon aeration, moisture content, temperature and available nutrients. The rate of decomposition is reduced when aeration is limited, under low or excessive moisture conditions, with low temperatures, and when nutrients, primarily nitrogen, are limited.

Within six months most of the celluloses disappear, but the lignins may last in soil for years. Examples of some organic residues and their relative decomposition rates are given below:

Grass clippings
Manures   } . . . . . . Rapid — days to weeks
Mushroom compost

Leaf mold
Composts $\Big\}$ . . . . . . Up to about six months or more
Humus-type compost

Rice hulls
Redwood
Fir bark $\Big\}$ . . . . Longer-lasting — possibly for years
Cedar
Cypress

TYPES OF PHYSICAL SOIL AMENDMENTS

Physical soil amendments are classified into two broad categories: organic and inorganic. Organic amendments are derived from living sources. They improve soil properties by physically separating the soil particles and by increasing nutrient- and water-holding capacity. The stable decomposition residue from organic amendments is humus.

Inorganic soil amendments do not contain humus nor do they contribute to the production of it. Instead, their principal function is to act as wedges in separating the soil particles physically. Some of these materials also aid in the retention of water.

A description of the more common organic and inorganic amendments follows.

Organic Amendments

Peat — Peat is composed of plant residues that have accumulated and undergone partial decomposition in water or in excessively wet areas such as swamps and bogs. Peats and peat moss exhibit a considerable range in moisture-holding capacity, organic content and longevity. Once they become dry, peats are difficult to wet.

Peats are classified into different types, according to the kinds of plants from which they were derived and the length of time they have been decomposing.

Moss peat (peat moss, sphagnum peat) — This partially decomposed peat is very low in plant nutrient content and has an acid pH. It has a high water-holding capacity, equal to 15 to 30 times its own weight. Moss peat is lightweight, porous and somewhat difficult to mix into soils. It is light brown to tan in color. Moss peat decomposes at a moderate rate because of its high cellulose content.

Fibrous peat (reed-sedge) — This peat has the highest commercial value and is of the most interest to turfgrass and landscape people. It accumulates in swamps or bogs along the edges and beneath the moss layers in the bogs. Reddish to dark brown in color, this rather fibrous material has undergone longer decomposition and is resistant to further decay. It does not have as high a moisture-holding capacity as moss peat.

Wood sawdust and shavings — Various grades of wood residues are widely used in amending turfgrass and landscape soils and in preparing container soil mixes. While these amendments provide good water infiltration and oxygen diffusion into soils, they possess only poor to fair water- and nutrient-holding capacities.

The rate of decomposition varies with the kind of wood used. Residues from pine and fir may have a residual of only several months, while residues from redwood, cedar and cypress may last up to five years. When these materials are incorporated into a soil mixture, additional nitrogen must be added to prevent nitrogen starvation of plants growing in the amended soils.

Ground fir bark — This product has proven to be a valuable amendment in the turfgrass and landscape industry. It has low nutrient-holding and fair water-holding capacity. The round shape of the bark provides good water and oxygen diffusion rates into the soil and resists compaction. Ground fir bark lasts about five years in a soil mix.

Fig. 9-4. Bulk redwood soil amendment in storage.

Shredded rice hulls — Rice hulls are the seed coats of rice. They are excellent conditioners, since they hold soils open and increase soil porosity. Rice hulls, however, have neither a good water- nor a good nutrient-holding capacity. The hulls decay very slowly in a soil mix, lasting up to 10 years.

Composts — Composts are made by the decomposition of plant residues. They usually have a good nutrient-holding capacity and high biological activity. Well-decomposed composts aid in creating desirable soil structure. One special type is mushroom compost, a by-product of the mushroom growing beds, consisting of soil, straw, manure, wood or peat, and agricultural minerals. It contributes moderate nutrients and low humus to soil mixes.

Composted manures — Steer and dairy manures are best used when they have been well aged and composted. This reduces the problems of odor and weed seed contamination. The water- and nutrient-holding capacities of composted manures are fair.

Digested sludges — These sludges from sewage processing are produced after undergoing about 14 days of aerobic (with air) and anaerobic (without air) digestion. Afterwards, they are pumped into basins and allowed to dry. Unless modified by the addition of wood products, digested sludges are not generally suitable as a soil amendment. Perhaps their best value is realized when they are used as an organic base for fertilizers.

Fig. 9-5. A golf green showing the value of an organic amendment (sludge) in the foreground vs. pure sand in the background.

Composted sludges — This product is made by centrifuging digested sludge to reduce water content. It is then windrowed with biologically active dried sludge and turned daily for about 40 days. The resulting well-decomposed product approaches the characteristics of humus. When used in silt and clay soils, it improves aggregation; when used in sandy soils, it improves moisture-holding capacity and biological activity.

Organically Amended Soil Mixes

These are sometimes referred to as *artificial soils* and have experienced a large increase in popularity in recent years. They developed on a commercial scale in the mid-1950's as a consequence of an overabundance of wood by-products from logging operations. With the growth of the container and landscape industry has come a general shortage of the more desirable amendments, mainly redwood bark and sawdust. Organically amended soil mixes have become an integral part of controlled plant culture. Their uses extend from growing hothouse vegetables to producing nursery plants, and building golf greens and athletic fields. Homeowners purchase these mixes for house and patio plants.

There are many reasons for using organically amended soil mixes. Often they provide a superior medium for producing quality plants. The use of turf and ornamental plants in roof gardens and above-ground patios necessitates low-density growing media. Low-density soils reduce transportation costs.

In choosing organic soil amendments, the first consideration should be to use materials resistant to decomposition. Redwood bark and sawdust, as well as the bark of fir and pine, are the more resistant wood by-products.

Moisture retention and pore space are important considerations when choosing among various amendments. Percent moisture retention is usually given as volume percent for convenience. For instance, if redwood sawdust has a moisture retention capacity of 50 percent, then a 1-gallon can of sawdust holds ½ gallon of water. It is desirable to use products which hold large volumes of water so long as there is sufficient air present. Total porosity and air accessibility are two very important physical properties of any growing medium.

Moisture–air relationships are presented in Figure 9-6. Following irrigation and drainage (when the medium is at container capacity) a perched water table is present at the bottom of the container. The moisture content here will be at saturation.

The physical properties of different media vary widely, as illustrated in Table 9-2.

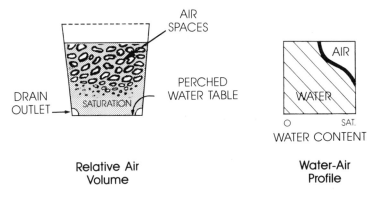

Fig. 9-6. Moisture-air relationships in containers at "container or field capacity."

Table 9-2

Physical Properties of Two Container Media at Container Capacity[1]

	Percent by Volume				
Medium	Total Porosity	Air Porosity	Water Content	Dry Density (g / cc)	Wet Density (lbs. / cu ft.)
1 Sand + 2 Peat	74	10	64	0.64	80
1 Perlite + 1 Peat	94	20	74	0.13	54

[1]Depth of medium = 5 inches.

Media requirements of a wide variety of ornamentals are given in Table 9-3.

Organic soil amendments do not have high nutrient-supplying capacities and, therefore, amended soil mixes require abundant quantities of the nutrient elements, particularly nitrogen. Consequently, many commercial nursery operations are on a constant feeding program.

Synthetic soils may contain from 50 to 100 percent organic amendments. Typical soil mixes prepared for landscaping, golf greens, nursery containers, bedding plants, cut flower beds and athletic field turf consist of 50 percent organic amendment mixed with fine sand or loam soil plus nutritive and pH-adjusting minerals. The organic amendment is normally red-

Table 9-3

Approximate Root Aeration Requirements of Selected Ornamentals Expressed as Percent Air Space

Very High 20	High 20 – 10	Intermediate 10 – 5	Low 5 – 2
Azalea	African violet	Camellia	Carnation
Fern	Begonia	Chrysanthemum	Conifers
Orchid, epiphytic	Daphne	Gladiolus	Geranium
	Foliage plants	Hydrangea	Ivy
	Gardenia	Lily	Palm
	Gloxinia	Poinsettia	Rose
	Heather		Stocks
	Orchid, terrestrial		Strelitzia
	Podocarpus		Turf
	Rhododendron		
	Snapdragon		

wood or fir bark or sawdust, fortified with fertilizer. Sphagnum peat moss is often a part of the organic component because of its high moisture retention qualities.

Potting soils and soils for landscape plantings contain at least two-thirds organic amendment mixed with the same constituents as previously listed. Specialty crops such as azaleas, rhododendrons, heather and orchids are commercially grown in 100 percent organic soil mixes. The approximate number of containers required to utilize 1 cubic yard of an organically amended soil mix is given in Table 9-4.

Table 9-4

Approximate Number of Containers Filled from 1 Cubic Yard of Organically Amended Soil Mix[1]

300 one-gallon cans	1,500 four-inch pots
500 six-inch pots	4,000 three-inch pots
800 five-inch pots	

Three cu yds. per 1,000 sq. ft. will provide 1 in. of depth.

[1]Source: Used with permission of O. A. Matkin.

Formulations for particular requirements should be obtained from commercial consultants or university horticultural advisors. There are also numerous publications available with more specific information on this subject.

Inorganic Amendments

Calcined clay — This long-lasting fired clay resists compaction very well and promotes water infiltration and oxygen diffusion. Water-holding capacity is high, but water-releasing and nutrient-holding characteristics are poor. It makes a good surface mulch. In clay soils, it helps to develop very deep roots, but with only a sparse density. While costly, these materials can aid in the establishment of high-value turfgrass and landscape plantings. Soils amended with calcined clays often have a hard surface.

Pumice — This amending material is a spongy, light, porous volcanic rock. It is relatively costly but has good properties in helping to establish plantings. When mixed into the soil, it promotes good water infiltration and oxygen diffusion rates and a deep, dense root system. Water- and nutrient-holding properties of pumice are fair to poor. Pumice from some sources may have excessively fine pores, which results in poor water-releasing properties. This material is usually limited to use in greenhouse propagation mixes and specialized potting soils.

Vermiculite — Upon heating a naturally occurring micaceous mineral ore, the internal moisture turns to steam and pops the flakes to many times their original size. The resultant fluffy, lightweight product can hold water in amounts several times its own weight, although even when thoroughly wet, there is still sufficient air (oxygen) for plant roots to grow. Lasting indefinitely, vermiculite is used mostly in specialized greenhouse propagation and potting soil mixes.

Perlite — Perlite is a naturally occurring, glassy volcanic rock which is heat-treated like vermiculite to produce a whitish-colored, bead-like particle with very low bulk density. It is used as a soil amendment to provide good aeration and drainage. Perlite is often used in combination with peat moss as a medium for the propagation of ornamental nursery stock and as a lightweight ingredient in potting soils for both commercial and home use. Perlite is inert and water-insoluble. It has a pH of 6 to 7.

Sand — This low-cost, long-lasting amendment is used in special sand mixes such as on golf and bowling greens and in nursery production. Sand is low in both water- and nutrient-holding capacity and causes finer silt or clay soils to compact. It thus has limited value as a soil amendment unless used at very high rates, perhaps up to 80 percent of the mix.

SELECTING PHYSICAL SOIL AMENDMENTS FOR VARYING SOIL CONDITIONS

Table 9-5 presents the soil-amending properties of some commonly used amendments. No single material may be totally satisfactory, so a combination of amendments is often used. Sandy soils are benefited when amended with a humus type of material to increase water retention and nutrient-holding capacity. Clayey and silty soils, in contrast, need fibrous-type materials to increase permeability and water retention. Humus may be needed to give better water and nutrient retention if the soil mix is low in organic matter content.

Table 9-5

Soil-amending Properties of Some Materials

	Amendment	Permeability	Water Retention
Fibrous	Peat	Low-Medium	Very high
	Wood residues	High	Low-Medium
	Ground fir bark	High	Low-Medium
	Rice hulls	High	Low-Medium
Non-fibrous or humus	Composts	Low-Medium	Medium-High
	Composted manures	Low-Medium	Medium
	Composted sludges	Low	High
Inorganic	Calcined clay	High	High
	Pumice	High	Low-Medium
	Vermiculite	High	High
	Perlite	High	Low
	Sand	High	Low

DETERMINING QUANTITIES OF ORGANIC AMENDMENTS

When choosing amendments, consider (1) whether they will be effective, practical and economical and (2) the use, size and value of the area to be amended. A good loam soil not subject to much compaction may not need amending. It is generally ineffective to amend a soil using less than 25 percent by volume of amendment. Heavy-use, high-value areas and fine textured soils with poor structure may benefit from being amended up to

Table 9-6

Amendment Needed, Based on Soil Texture

Texture	Percent of Amendment
Sand	35
Loamy sand Sandy loam	30
Sandy clay loam Sandy clay Loam	25
Silt loam Silty clay loam Clay loam	30
Silt Silty clay Clay	35

40 or 50 percent by volume. Some general guidelines in determining the quantity of soil amendments needed for various soil textures are given in Table 9-6.

Turfgrass and Ground Covers

The general rule to amend soils for these plants is to use about 33 percent by volume of amendment. For heavily used areas and with fine textured silt and clay soils, this may be increased to 40 percent by volume. For good loam soils where little traffic is expected, this may be reduced to a minimum of 25 percent by volume. Table 9-7 may be used to determine the volume of soil amendment needed to amend the soil to various depths.

Bedding Plants, Ground Covers and Vegetable Gardens

These are usually high-volume, low-acreage areas which may be amended by adding 2 inches of amendment over a previously deeply cultivated soil. If the grower incorporates the amendment to a 6-inch depth, the soil will be amended 33 percent by volume.

Table 9-7

Determining the Volume of Soil Amendments to Add for Various Depths of Treatment (cubic yards per 1,000 square feet)[1]

Percent of Amendment	Depth of Amended Soil in Inches (Soil and Amendments)						
	3 in.	4 in.	5 in.	6 in.	7 in.	8 in.	9 in.
1	0.09	0.12	0.15	0.18	0.21	0.25	0.28
2	0.18	0.25	0.31	0.37	0.43	0.49	0.55
3	0.28	0.37	0.46	0.56	0.65	0.74	0.83
4	0.37	0.49	0.62	0.74	0.86	0.99	1.11
5	0.46	0.61	0.77	0.93	1.08	1.23	1.39
10	0.93	1.23	1.54	1.85	2.16	2.47	2.78
20	1.85	2.47	3.09	3.71	4.32	4.94	5.55
30	2.78	3.70	4.64	5.56	6.48	7.41	8.33
40	3.70	4.94	6.18	7.41	8.64	9.88	11.13
50	4.63	6.17	7.72	9.26	10.80	12.34	13.88

[1]These values are additive; i.e., to calculate values not in the table, simply add the percentage figures. For example, to calculate 33% at a 6-inch depth, add the 30% figure and the 3% figure (5.56 + 0.56 = 6.12).

Backfill Around Landscape Shrubs and Trees

Recommendations are based on the size of the planting hole. Table 9-8 indicates some hole sizes for various containers and the quantities of amendments to be added to provide either 33 percent or 50 percent by volume.

Container Mixes

Soil mixes for container plantings vary widely in composition, depending upon use and availability of materials. The following two examples are formulas which have given good success, although they are often modified.

Example 1 — Landscape container mix

50% — sand or sandy loam

50% — fibrous organic material such as redwood or rice hulls

plus, per each cubic yard add:

4 oz. potassium sulfate

6 oz. potassium nitrate

Table 9-8

Determining the Volume of Soil Amendments (cubic feet) Required to Amend Backfill Soils by 33% or 50% for Various-sized Containers (cans or boxes)

		Container Size									
	1 Gal.		5 Gal.		7 Gal.		15 Gal.		20 In.		24 In.	
Hole Size[1]	33%	50%	33%	50%	33%	50%	33%	50%	33%	50%	33%	50%
2W, D	0.1	0.2	0.8	1.2	1.3	2.0	1.6	2.4	3.8	5.7	6.7	10
2W, D + ½ foot	0.3	0.5	1.3	2.0	2.0	3.0	2.6	4.0	5.4	8.2	9.2	14
2W, D + 1 foot	0.4	0.6	1.8	2.7	2.7	4.1	3.5	5.3	6.3	11.0	12.0	17
2W, 2D	0.3	0.5	1.8	2.7	2.9	4.4	3.6	5.5	8.8	13.0	16.0	24

	30 In.		36 In.		42 In.		48 In.		54 In.		60 In.	
Hole Size[1]	33%	50%	33%	50%	33%	50%	33%	50%	33%	50%	33%	50%
2W, D	14	22	23	34	30	45	44	67	52	79	73	111
2W, D + ½ foot	18	27	28	43	37	56	54	82	64	97	87	132
2W, D + 1 foot	23	39	34	51	44	67	64	97	75	114	101	154
2W, 2D	33	50	53	80	69	105	103	156	121	184	172	260

[1]W = width of container; D = depth of container.

2½ lbs. single superphosphate
3½ lbs. dolomite lime
1½ lbs. calcium carbonate – lime
1¼ lbs. gypsum

Example 2 — Greenhouse or foliage plant mix

33% sand
33% peat moss
33% redwood or perlite

plus, per each cubic yard add:
2 lbs. single superphosphate
4 oz. potassium sulfate
4 oz. potassium nitrate
5 lbs. hoof and horn meal or blood meal

INCORPORATION OF SOIL AMENDMENTS

A uniform soil or container mixture must be provided throughout the entire root zone. Poor mixing of materials will produce layers or pockets,

Fig. 9-7. Bulk organic soil amendments being mixed for container plants.

resulting in interfaces that can restrict air, water, fertilizer and root movement through the soil.

With the exception of potting soils or specialty growing media, physical soil amendments are mixed with the existing site soil. This is described as on-site mixing. On-site mixing involves the uniform distribution of the desired materials in layers over the surface of the soil followed by thorough mixing with cultivating equipment, ensuring that no layers or pockets remain. Rototillers, turning cultivators, discs and harrows are often used for mixing. Note: Discs tend to leave pockets at turns, and high-speed tillers may destroy soil structure.

After cultivation, the area should be graded to provide for positive drainage, while ensuring that no low spots or pockets remain.

Potting soils or specialty media are usually mixed in concrete-mixer type blenders. They are often steam-sterilized before using.

Mulches

Mulches are those materials that are spread upon the ground to:

1. Retain soil moisture.
2. Protect against soil erosion.
3. Protect the roots of plants from heat, cold or drought.
4. Keep fruit clean.
5. Allow earlier planting.
6. Assist in providing uniform seed germination and plant establishment.

Not only do mulches control soil erosion that causes gullies and rills, but they also protect against sheet erosion caused by the action of splashing raindrops. They also prevent the displacement of seeds and fertilizers and can aid in minimizing weeds and surface soil crusting.

Commonly used mulch materials are straw, plastic, wood chips or bark, sludge and fibrous cellulose materials such as those used in hydroseeding. Combinations of these products are also used.

Mulching and Top Dressing

Generally, mulching involves the spreading of materials at depths of ½ inch or greater (Table 9-9). Top dressing should be limited to depths of ⅛ inch to ¼ inch. Applying materials thicker than this after seeding of lawns or ground covers may interfere with germination and water movement and promote fungal disease. On such areas it would be wise to use the faster-

Fig. 9-8. Nurseryman inspecting root growth of plants grown in prepared organic medium.

Table 9-9

Volume of Materials Required for Mulching or Top Dressing

Depth (in.)	Approximate Volumes of Material	
	Per 1,000 Sq. Ft. (cu. ft.)	Per Acre (cu. yds.)
⅛	10	16
¼	20	32
⅜	30	48
½	40	64
⅝	50	80
¾	60	96
⅞	70	112
1	80	128

decomposing materials; those with longer residual duration may contribute to thatch and eventual water infiltration problems.

By necessity, this chapter has been concerned only with the physical properties of soils and soil amendments. To explore this subject more fully, examine the supplementary reading references. Additionally, since the organic soil amendments have other chemical and biological values, read Chapters 1 and 8 for a more complete understanding of these materials.

SUPPLEMENTARY READING

1. *Amending Soils.* P. A. Rogers. California Landscape Management. March / April 1976.
2. *Ball Red Book,* Fourteenth Edition. V. Ball, ed. Reston Publishing Co. 1984.
3. "Evaluating Soil Amendments." J. E. Warneke and S. J. Richards. *California Agriculture.* September 1974.
4. "Peat Classifications." J. R. Watson. *California Turfgrass Culture.* July 1967. *Turfgrass Times.* 1968.
5. "Review of Soil Amendments." W. H. McNeely and W. C. Morgan. *Turfgrass Times.* March 1968.
6. "Soil Amendments, What Can They Really Do?" R. L. Branson. *Western Landscape News.* February 1979.
7. *Soil Fertility and Fertilizers,* Fourth Edition. S. L. Tisdale, W. L. Nelson and J. D. Beaton. The Macmillan Company. 1985.
8. *Soil Preparation Specifications.* Kellogg Supply, Inc. 1978.
9. "The Use of Physical Soil Amendments in Turfgrass Management." W. C. Morgan, J. Letey, S. J. Richards and N. Valoras. *California Turfgrass Culture.* July 1968.

CHAPTER 10

Soil/Media and Tissue Testing

Soil and plant tissue analyses are the horticulturist's best guide to the wise and efficient use of fertilizers and soil amendments. They are a primary factor in a high quality, maximum profit, best management practices program. Useful recommendations from soil and tissue tests require accurate sampling, analysis and interpretation based on sound research, practical experience and judgment. The interpretive guides presented here are those that apply generally in the West. The user is advised to contact the local agricultural extension service, experiment station or qualified industry representatives for recommendations for plantings in specific areas. These guides should be used only with data obtained from samples collected and analyzed by the procedures specified.

Each diagnostic technique has both advantages and limitations. Soil and plant analyses do different jobs and should be used in such a way that they support and supplement each other. Soil analyses are most useful in appraising nutrient requirements and in evaluating pH and salt problems in soils and planting media. They have the advantage that they can be completed and used prior to planting.

Plant analyses are particularly useful in determining the nutritional status of established, deep-rooted ornamentals such as trees, shrubs and vines, where soil samples of the entire feeding zone are difficult to obtain and interpret. They are also useful in diagnosing the causes of poor growth, in evaluating the effectiveness of fertilizer treatments, in following the nutrient status of plants throughout the growing season and in managing quality factors.

Satisfactory recommendations based on soil or tissue tests depend upon three factors: representative sampling, accurate analysis and proper interpretation of the analytical results.

SAMPLING

The sample must correctly represent the area of the soil, or the crop being sampled. The results obtained can be no better than the sample collected and analyzed. No single set of sampling instructions can apply to all · situations.

The proper method for collecting and handling samples is determined by:

1. The use to be made of the analyses.
2. The pattern and ease of recognition of soil or crop variability.
3. Previous and proposed management practices.

Generally, soils and media mixes used in horticultural plantings are more homogeneous than native agricultural soils. But variation can be great between plant beds, media mixes and surface amended soils. It is important to recognize this and, if practical, measure it. An average test value from a highly variable area or soil mix may be of little use. In sampling an area, one should first divide it into relatively homogeneous and manageable units. The divisions can be based on the appearance of the crop or soil surface, management history, drainage, texture and erosion. A composite sample of at least 10, and preferably 20, subsamples should be collected from each unit. To diagnose the causes of poor or abnormal growth, individual samples from adjacent normal and affected areas should be collected.

The depth to which samples should be taken is determined by the species grown, the proposed use of the analysis and what is already known about the medium. Evaluation of soluble salt levels and of levels of nutrients that move freely within the profile requires samples taken through the rooting depth. Deep samples are frequently of value in explaining unexpected growth patterns resulting from either chemical or physical characteristics of subsurface layers. For many uses, samples taken to tillage depth are adequate; however, one should take separate samples at the surface and subsurface if sodium or salinity problems are anticipated.

The time to take soil or media samples is determined by the information desired. Samples for determining fertilizer needs of annuals should be taken sufficiently in advance of planting to allow time for analysis and return of the results from the laboratory. Samples for diagnosing the cause of poor plant growth or for evaluating salt or sodium hazards are best taken when the problem areas are delineated by plant or other visual differences.

In sampling plant tissues, the species, its stage of growth, the part of the plant to be analyzed and variation within the block must be considered. The nutritional status of most perennials, e.g., trees, vines, shrubs and turfgrass, may be evaluated from samples taken at a single growth stage. For annuals and perennials in nurseries, it is better to make samplings at several stages of growth so that impending nutrient deficiencies can be detected and corrected before severe quality and / or yield reductions occur.

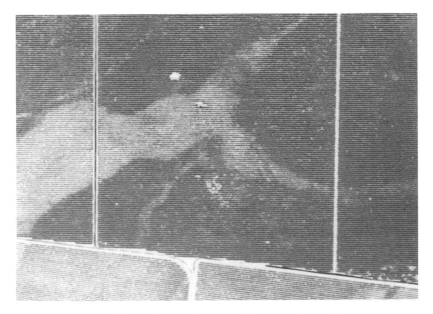

Fig. 10-1. Aerial view of a vineyard showing soil variability.

Interpretation of these analyses is based on guides established from analyses of specific tissues sampled at specified times. For this reason the plant parts collected must correspond with those of the interpretive guide used. Plants, like the media on which they are growing, are not uniform over large areas, and for this reason a block to be sampled should first be subdivided into uniform soil / soil mix and cropping areas. Thirty to 50 plant parts are then collected as the representative sample from an area.

Before submitting either soil or plant tissue samples for analysis, one should:

1. Obtain containers, instructions and information sheets from the laboratory.
2. Follow instructions carefully in collecting and preparing the samples.
3. Keep accurate records of the areas sampled, including fertilizers, amendments, pesticides and crop histories, for future reference.

ANALYTICAL METHODS

It is necessary to choose the kinds of analyses that are meaningful when the sample is sent to the laboratory, since laboratories can analyze for many things. Also, the results of these analyses are useful only insofar as the methods are reliable and the data interpretable. Soil analyses of general use in evaluating levels of available nutrients in media containing western soils include pH (measured in a soil-water paste), $NaHCO_3$-extractable P (or other methods for acid soils), exchangeable K (neutral salt extraction) and DTPA-extractable Zn. Detailed methods for these determinations are available from several sources (see Supplementary Reading).

Plant tissue analyses are used to identify inadequate, normal or excessive amounts of several elements. Thus procedures for a wide range of measurements are required. The most common determinations include the concentration of nitrate ($NO_3 - N$), total or Kjeldahl N, 2 percent acetic acid – soluble phosphate ($PO_4 - P$), total P, readily reducible sulfate ($SO_4 - S$) and the total contents of K, Ca, Mg, S and the micronutrients. Suitable methods for these measurements are available from several sources (see Supplementary Reading).

The results of nutrient analyses are most often reported as the concentration of the element in oven-dry soil or plant tissue. Some laboratories report the results of soil / media analyses on the air-dry basis.

INTERPRETIVE GUIDES

Translation of soil and tissue test data into meaningful management recommendations involves the integration of all available pertinent information. It should be done by experienced horticulturists who understand the principles of fertility and plant nutrition and who are familiar with production conditions and practices, and fertilizer response data in the area.

Soil Tests

Guides to the interpretation of soil tests for available P, K and Zn are in Table 10-1. The values listed, although generally applicable, cannot be applied to all situations. Bicarbonate-extractable P is not a reliable index of phosphate availability in strongly acid or high-organic-matter mixes. Other soil tests are used in these situations. Similarly, interpretations of exchangeable K levels should take into account the rate of release of nonexchangeable K, subsoil K levels and the influences of levels of other cations. In spite of their limitations, these tests provide growers with estimates of fertilizer needs of their soils and media at the time the information is needed most and is not available from other sources.

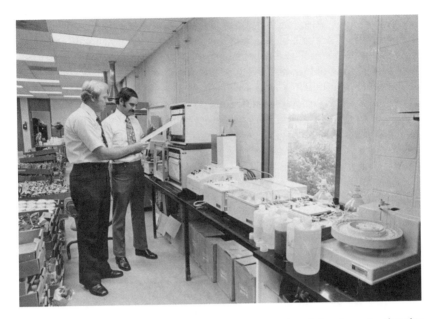

Fig. 10-2. Modern soil and tissue laboratories contain automated equipment such as this Multichannel Technicon Auto Analyser II Continuous Flow Analytical System.

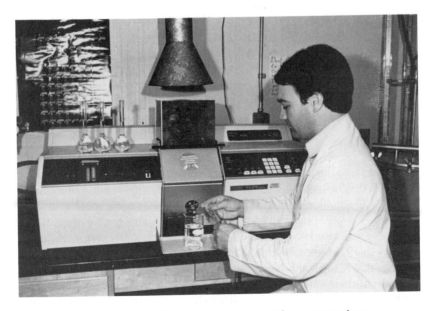

Fig. 10-3. Modern laboratory equipment provides precise analyses.

Soil pH provides a principal clue in the diagnosis of many soil/media problems. Characteristics such as the solubility and availability to plants of several important nutrient elements, the rates of microbial reactions and the level of exchangeable sodium are closely related to the pH of the soil. In strongly acid media (pH below 5.5), the solubility of aluminum and manganese and the heavy metals is high, and yields and quality frequently suffer due to toxic levels of one or more of these elements. At the other end of the scale are the strongly alkaline substrates (pH above 8.5); these contain precipitated carbonates (calcareous soils) and in some cases, depending on

Table 10-1

Generalized Soil Test Interpretive Guide

Most Field and Greenhouse Plants	$NaHCO_3 - P$[1]	Exchangeable K	DTPA Zn
	(ppm)		
Highly responsive	<8	<60	<0.5
Probably responsive	8 – 15	60 – 100	0.5 – 1.0
Not responsive	>15	>100	>1.0

	$NaHCO_3 - P$		Exchangeable	DTPA
Potatoes	Pac. N.W.	S.W.	K[2]	Zn
	(ppm)			
Highly responsive	<10	<12	<100	<0.5
Probably responsive	10 – 20	12 – 25	100 – 150	0.5 – 0.8
Not responsive	>20	>25	>150	>0.8

[1]For cool season crops, increase these values 30%.
[2]Exchangeable K values 50% to 100% higher are suggested for Washington and Oregon.

irrigation regime, appreciable amounts of exchangeable sodium. Plant growth in these situations may be limited by inadequate supplies of iron (lime-induced chlorosis) and zinc, or by sodium problems.

Most unamended horticultural soils are in the pH range of 5.5 to 8.0. Barring the overriding effects of sterilization, the growth of plants on these soils is influenced by the favorable effects of near-neutral reaction on nitrification, symbiotic nitrogen fixation and availability of plant nutrients. The optimum pH range for most species is 6.0 to 7.5 and for legumes, 6.5 to 8.0.

Tissue Tests

Guides to the interpretation of tissue tests for principal western ornamentals are given in Tables 10-2 to 10-6. Nutrient levels are given in relation to the probability of response from application of fertilizers supplying the element. Tissue test values at or below the "deficient" levels indicate a high probability of response; those above "sufficient" indicate that adequate supplies are available. Response may or may not be obtained with tissue levels between "deficient" and "sufficient," depending on production levels and other factors. Concentrations exceeding the "excess" levels may be associated with reduced growth or quality. It is very important that samples of the same plant part, taken at the same physiological age, be analyzed if the data presented here are to be used as a guide.

As an example of using tissue tests to guide a fertilizer program, the influence of leaf levels of nitrogen and potassium on yield and quality factors of oranges is presented in Figures 10-4 and 10-5. Increasing the level of a given element in the tree influences some factors favorably and others unfavorably. In most years, benefits are maximum when the nitrogen and potassium concentrations are in the optimum ranges.

Table 10-2

Plant Tissue Analysis Guide for Western Crops

Plant	Time of Sampling	Plant Part		Nutrient Level[1]	
				Deficient	Sufficient
Asparagus	Midgrowth of fern	4" tip section of new fern branch	N P K	100 800 1	500 1,600 3
Bean, bush snap	Midgrowth	Petiole of fourth leaf from growing tip	N P K	2,000 1,000 3	4,000 3,000 5
Bean, bush snap	Early bloom	Petiole of fourth leaf from growing tip	N P K	1,000 800 2	2,000 2,000 4
Broccoli	Midgrowth	Mid-rib of young, mature leaf	N P K	7,000 2,500 3	10,000 5,000 5
Broccoli	First buds	Mid-rib of young, mature leaf	N P K	5,000 2,000 2	9,000 4,000 4

(Continued)

Table 10-2 (Continued)

Plant	Time of Sampling	Plant Part		Nutrient Level[1] Deficient	Sufficient
Brussels sprouts	Midgrowth	Mid-rib of young, mature leaf	N P K	5,000 2,000 3	9,000 3,500 5
Brussels sprouts	Late growth	Mid-rib of young, mature leaf	N P K	2,000 1,000 2	4,000 3,000 4
Cabbage	At heading	Mid-rib of wrapper leaf	N P K	5,000 2,500 2	9,000 3,500 4
Cantaloupe	Early growth (short runners)	Petiole of sixth leaf from growing tip	N P K	8,000 2,000 4	12,000 4,000 6
Cantaloupe	Early fruit	Petiole of sixth leaf from growing tip	N P K	5,000 1,500 2	9,000 2,500 4
Cantaloupe	First mature fruit	Petiole of sixth leaf from growing tip	N P K	2,000 1,000 2	4,000 2,000 4
Carrot	Midgrowth	Petiole of young, mature leaf	N P K	5,000 2,000 4	10,000 4,000 6
Cauliflower	Buttoning	Mid-rib of young, mature leaf	N P K	5,000 2,500 2	9,000 3,500 4
Celery	Midgrowth	Petiole of newest fully elongated leaf	N P K	5,000 2,000 4	9,000 4,000 7
Celery	Near maturity	Petiole of newest fully elongated leaf	N P K	4,000 2,000 3	6,000 4,000 5
Cucumber (pickling)	Early fruit-set	Petiole of sixth leaf from growing tip	N P K	5,000 1,500 3	9,000 2,500 5
Lettuce	At heading	Mid-rib of wrapper leaf	N P K	4,000 2,000 2	8,000[2] 4,000 4
Lettuce	At harvest	Mid-rib of wrapper leaf	N P K	3,000 1,500 1.5	6,000[2] 2,500 2.5
Pepper, chili	Early growth	Petiole of young, mature leaf	N P K	5,000 2,000 4	7,000 3,000 6

(Continued)

Table 10-2 (Continued)

Plant	Time of Sampling	Plant Part		Nutrient Level[1]	
				Deficient	Sufficient
Pepper, chili	Early fruit-set	Petiole of young, mature leaf	N	1,000	2,000
			P	1,500	2,500
			K	3	5
Pepper, sweet	Early growth	Petiole of young, mature leaf	N	8,000	12,000
			P	2,000	4,000
			K	4	6
Pepper, sweet	Early fruit-set	Petiole of young, mature leaf	N	3,000	5,000
			P	1,500	2,500
			K	3	5
Potatoes	Early season	Petiole of fourth leaf from growing tip	N	8,000	12,000[3]
			P	1,200	2,000
			K	9	11
Potatoes	Midseason	Petiole of fourth leaf from growing tip	N	6,000	9,000[3]
			P	800	1,600
			K	7	9
Potatoes	Late season	Petiole of fourth leaf from growing tip	N	3,000	6,000
			P	500	1,000
			K	4	6
Rose clover	Flowering	Leaves	N	—	—
			P	1,200	1,500
			K	0.7	1.0
			S	130	180
Spinach	Midgrowth	Petiole of young, mature leaf	N	4,000	8,000
			P	2,000	4,000
			K	2	4
Subclover	Third flower	Fully expanded leaves	N	—	—
			P	800	1,000[4]
			K	0.7	1.0
			S	150	200
Sweet corn	Tasseling	Mid-rib of first leaf above primary ear	N	500	1,500
			P	500	1,000
			K	2	4
Sweet potato	Midgrowth	Petiole of sixth leaf from growing tip	N	1,500	3,500
			P	1,000	2,000
			K	3	5
Tomato (canning)	Early bloom	Petiole of fourth leaf from growing tip	N	8,000	12,000
			P	2,000	3,000
			K	3	6
Tomato (canning)	Fruit 1" diameter	Petiole of fourth leaf from growing tip	N	6,000	10,000
			P	2,000	3,000
			K	2	4

(Continued)

Table 10-2 (Continued)

Plant	Time of Sampling	Plant Part		Nutrient Level[1] Deficient	Sufficient
Tomato	First color	Petiole of fourth leaf	N	2,000	4,000
(canning)		from growing tip	P	2,000	3,000
			K	1	3
Watermelon	Early fruit	Petiole of sixth leaf from	N	5,000	9,000
		growing tip	P	1,500	2,500
			K	3	5

[1]Unless otherwise noted, values are: N = NO_3 – N, ppm; P = acetic acid – soluble PO_4 – P, ppm; K = total K, %; and S = SO_4 – S, ppm.

[2]Nitrate concentrations 30% higher are suggested for winter-grown lettuce in the desert valleys of Arizona and southern California.

[3]Nitrate levels 40% to 60% higher are suggested for potatoes growing in the high valleys of Idaho and Oregon. Approximately 60% of the total P of potato petioles is extracted with 2% acetic acid.

[4]The sufficient level of P for heavily grazed subclover may be 50% higher.

Other Tests

Other tests used include appraisals of salinity and alkali conditions; water quality; soil/media levels of other nutrient elements and organic matter; and soil moisture, texture, density and permeability to air and water. Methods of analysis and interpretive criteria for these determinations are presented in the Supplementary Reading publications.

Table 10-3

Interpretive Guide for Grape Tissue Analysis[1]

Nutrient	Deficient	Sufficient	Excess
NO_3 – N, ppm	<350	600 – 1,200	>2,400
P (total), %	<0.15	0.20 – 0.60	—
K (total), %	<1.0	1.50 – 2.50	>3.0
Mg (total), %	<0.3	0.5 – 0.8	>1.0
Zn (total), ppm	<15	25 – 50	—
B (total), ppm	<25	40 – 60	>300[2]
Cl (total), %	—	0.05 – 0.15	>0.50

[1]Concentrations of nutrients in Thompson seedless grape petioles collected opposite the cluster at full bloom. Adapted from information supplied by J. A. Cook.

[2]Concentration in leaf-blade tissue.

Table 10-4

Leaf Analysis Guide for Fruit and Nut Trees, Pacific Northwest[1]

		Level of Nutrient in Leaves[2]		
	Nutrient	Deficient Below	Satisfactory	Excess Above
Peach	N	2.5%	2.6% – 3.5%	3.5%
Sweet cherry	N	2.0%	2.0% – 3.0%	3.0%
Apricot, prune	N	2.0%	2.0% – 2.5%	3.0%
Filbert, walnut	N	2.0%	2.2% – 2.8%	2.8%
Apple, pear	N	1.7%	1.9% – 2.5%	2.5%
Apple, pear	K	0.9%	1.0% – 1.5%	—
Peach	K	1.0%	1.5% – 3.5%	—
Sweet cherry	K	1.0%	2.3% – 2.8%	—
Prune	K	1.0%	1.7% – 2.7%	—
Filbert, walnut	K	1.0%	1.2% – ?	—
Apple, pear	B	20 ppm	25 – 50 ppm	80 ppm
Stone fruits	B	20 ppm	35 – 80 ppm	100 ppm
All fruits	Zn	10 ppm	17 – ? ppm	—

[1]Adapted from Oregon State University FG 23, 24, 25, 26, 34 and 35; and Washington State University FG 28f and 28g.
[2]Midshoot leaves sampled between July 15 and August 15.

Table 10-5

Leaf Analysis Guide for Fruit
and Nut Trees, California[1]

	N[2] Ade-quate	K[3] Defi-cient	K[3] Ade-quate	Na[4] Excess	Cl[4] Excess	B Ade-quate	B Excess
		-------------------------------(%)-------------------------------				----------(ppm)----------	
Almond	2.0 – 2.5	1.0	1.4	0.25	0.3	30 – 65	85
Apple	2.0 – 2.4	1.0	1.2	—	0.3	25 – 70	100
Apricot (ship)	2.0 – 2.5	2.0	2.5	0.1	0.2	20 – 70	90
Apricot (can)	2.5 – 3.0	2.0	2.5	0.1	0.2	20 – 70	90
Sweet cherry	2.0 – 3.0	0.9	—	—	—	—	—
Fig	2.0 – 2.5	0.7	1.0	—	—	—	300
Olive	1.5 – 2.0	0.4	0.8	0.2	0.5	20 – 150	185
Nectarine and peach (freestone)	2.4 – 3.3	1.0	1.2	0.2	0.3	20 – 80	100
Peach (cling)	2.6 – 3.5	1.0	1.2	0.2	0.3	20 – 80	100
Pear	2.3 – 2.8	0.7	1.0	0.25	0.3	20 – 70	80
Plum (Japanese)	2.3 – 2.8	1.0	1.1	0.2	0.3	30 – 60	80
Prune	2.3 – 2.8	1.0	1.3	0.2	0.3	30 – 80	100
Walnut	2.2 – 3.2	0.9	1.2	0.1	0.3	26 – 200	300

Adequate levels for all fruit and nut crops: P, 0.1% – 0.3%; Cu, over 4 ppm; Mn, over 20 ppm; Zn, over 15 ppm.

[1]Leaves are July samples from nonfruiting spurs on spur-bearing trees, fully expanded basal shoot leaves on peaches and olives, and terminal leaflets on walnuts. Adapted from *Soil and Plant Tissue Testing in California*, James K. Beutel and Uriu and O. Lilleland, Univ. of California Ag. Sci. Div., Bull, No. 1879, April 1976, pp. 11–14.

[2]Percentage N in August and September samples can be 0.2%–0.3% lower than July samples and still be equivalent. Nitrogen levels higher than underlined values will adversely affect fruit quality and tree growth. Maximum N for Blenheims should be 3.0% and for Tiltons, 3.5%.

[3]Potassium levels between deficient and adequate are considered "low" and may cause reduced fruit sizes in some years. Potassium fertilizer applications are recommended for deficient orchards but test applications only for "low" K orchards.

[4]Excess Na or Cl causes reduced growth at the levels shown. Leaf burn may or may not occur when levels are higher. Salinity problems can be confirmed with soil or root analysis.

Table 10-6

Leaf Analysis Guide for Diagnosing Nutrient Status of Mature Valencia and Navel Orange Trees[1]

Element		Ranges[2]		
		Deficient Below	Optimum	Excess Above
N	%	2.2	2.4 to 2.6	2.8
P	%	0.09	0.12 to 0.16	0.3
K[3]	%	0.40	0.70 to 1.09	2.3?
Ca	%	1.6?	3.0 to 5.5	7.0?
Mg	%	0.16	0.26 to 0.6	1.2?
S	%	0.14	0.2 to 0.3	0.6
B	ppm	21	31 to 100	260
Fe[4]	ppm	36	60 to 120	250?
Mn[4]	ppm	16	25 to 200	1,000?
Zn[4]	ppm	16	25 to 100	300
Cu	ppm	3.6	5 to 16	22?
Mo	ppm	0.06	0.10 to 3.0	100?
Cl	%	?	<0.3	0.7
Na	%		<0.16	0.25
Li	ppm		<3	35?
As	ppm		<1	5
F	ppm		<1 to 20	100

[1]With the exception of N values, this guide can be applied to grapefruit, lemon and probably other commercial citrus varieties. Data of Embleton, Jones and Platt in *Soil and Plant-Tissue Testing in California,* 1978.

[2]Based on concentration of elements in 5- to 7-month-old, spring-cycle leaves from nonfruiting terminals. Leaves selected for analysis should be free of obvious tipburn, insect or disease injury, mechanical damage, etc., and be from trees that are not visibly affected by disease or other injury.

[3]Potassium ranges are for effects on number of fruits per tree.

[4]Leaves that have been sprayed with Fe, Mn or Zn materials may analyze high in these elements, but those of the next growth cycle may have values in the deficient range.

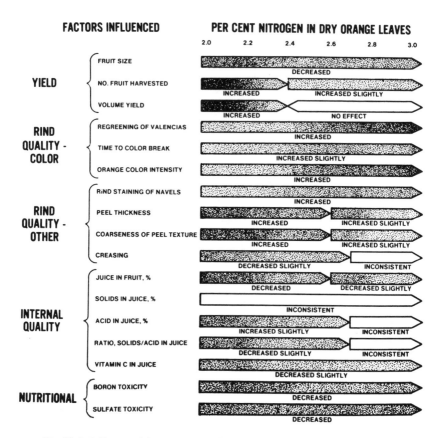

Fig. 10-4. Influence of the percentage of N in 5- to 7-month-old bloom-cycle leaves from nonfruiting shoots upon yield, and rind and fruit quality of oranges (adapted from Embleton, *et al.*, 1973).

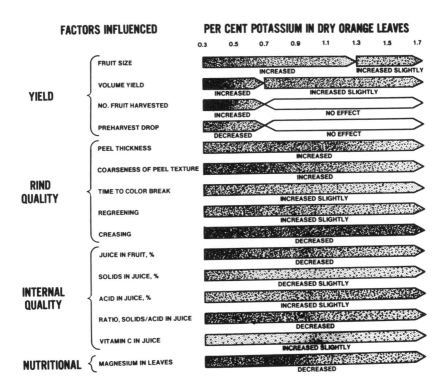

Fig. 10-5. Influence of the percentage of K in 5- to 7-month-old bloom-cycle leaves from nonfruiting shoots upon yield, and rind and fruit quality of oranges (adapted from Embleton, Jones and Platt in *Soil and Plant-Tissue Testing in California,* 1978).

SUPPLEMENTARY READING

1. *Analytical Methods for Use in Plant Analysis.* C. M. Johnson and A. Ulrich. California Agr. Expt. Bull. 766. 1959.
2. *Diagnosis and Improvement of Saline and Alkali Soils.* Agricultural Handbook No. 60. 1954.
3. *Diagnostic Criteria for Plants and Soils.* H. D. Chapman, ed. University of California, Division of Agricultural Sciences. 1966.
4. *Fertilizer Technology and Use,* Third Edition. O. P. Engelstad, ed. Soil Sci. Soc. of America. 1985.
5. *Hunger Signs in Crops,* Third Edition. H. B. Sprague, ed. David McKay Co., Inc. 1964.
6. "Leaf Analysis as a Diagnostic Tool and Guide to Fertilization." T. W. Embleton, W. W. Jones, C. K. Labanauskas and W. Reuther. *The Citrus Industry,* Vol. III (Revised). University of California, Division of Agricultural Sciences. 1973.
7. *Methods of Soil Analysis,* Part II. A. L. Page, ed. American Society of Agronomy. 1982.
8. *Soil and Plant-Tissue Testing in California.* H. M. Reisenauer, ed. University of California, Division of Agricultural Sciences. Bull. 1879. 1978.
9. *Soil Testing and Plant Analysis.* L. Walsh and J. D. Beaton, eds. Soil Sci. Soc. of America. 1973.
10. *Soil Testing Procedures for California.* CFA-SIC Publication. 1980.
11. *Sugarbeet Fertilization.* F. J. Hills, R. Sailsbery and A. Ulrich. University of California, Division of Agricultural Sciences. Bull. 1891. 1978.

CHAPTER 11

Methods of Applying Fertilizer

Fertilizers are used to supply nutrients that are not present in the soil or planting medium in amounts necessary to meet the needs of the growing plants. When choosing the methods of application, growers should consider the following:

1. The rooting characteristics of the species to be planted.
2. The plant's demand for various nutrients at different stages of growth.
3. The physical and chemical characteristics of the soil.
4. The physical and chemical characteristics of the fertilizer materials to be applied.
5. The availability of moisture.
6. The kind of irrigation systems used if irrigation is the only, or major, source of water.

Production of horticultural plants often requires multiple applications of fertilizer materials. Several methods of applying fertilizer may be employed. For example, established turfgrass may receive a spring and fall broadcast application of a complete fertilizer. To maintain turf quality and color, additional nitrogen is often applied broadcast or through the turf irrigation system.

Fertilizers added to soils undergo transformations that may change their availability. The methods of application are directly related to the plant's utilization of the nutrients and the changes the nutrients undergo in the soil. The application methods employed should be as economic, accurate and efficient as possible.

This chapter discusses different methods of fertilizer application and the equipment used in applying fertilizer. It suggests methods of adjusting the equipment for accurate applications and ways of overcoming some problems related to fertilizer applications.

Surface Application

The broadcast method of applying fertilizers consists of uniformly distributing dry or liquid materials over the soil surface.

Fig. 11-1. Drop spreader (left) and spinning-type spreader (right). Both are push types used by homeowners or for small areas.

Fig. 11-2. Turf being fertilized with typical spinning-type spreader.

Fig. 11-3. Liquid fertilizer storage and injection system feeding into sprinkler-irrigated golf course.

Drop spreader— The simplest applicator for dry fertilizer is an inverted triangle-shaped hopper mounted between two wheels. Fertilizer is distributed through adjustable openings in the bottom of the hopper. When properly adjusted or calibrated, very accurate application rates can be obtained.

Small bulk spreaders are available, each consisting of a bin mounted on a two- or four-wheeled trailer frame and pulled by a tractor, truck or ATV. The fertilizer is usually spread by a horizontal spinning disc in a 20- to 40-foot swath. The operator must exercise care to prevent skips or excessive overlaps of fertilizer.

Spinning-type spreader — The spreader consists of a bin mounted over a horizontal spinning disc (see Figure 11-2). Proper calibration of the spreader and ground speed permits a reasonably accurate and fast application. The operator must exercise care to prevent skips or excessive overlaps of fertilizer.

Liquid spreader — The basic components of a liquid fertilizer broadcast applicator are a tank, pressure gauge and regulator, pump, pipes, hoses, fittings, nozzles and a boom. The applicator can be mounted on a truck, flotation vehicle, ATV or trailer, or directly on a tractor. The speed of application is determined by the rate of flow.

Subsurface Application

Injection — "Injection" refers to placing fertilizers below the soil surface. Drop pipes for liquid fertilizers, or flexible tubes for dry fertilizers, deliver fertilizers into the channels made by the opening tools. All fertilizers that can be broadcast on the soil surface can also be injected. Because certain fertilizers are subject to losses when they are applied to the soil surface, many users choose to inject them.

Injection is also an excellent method of putting immobile nutrients into the root zone, where they may be more efficiently utilized. If soils are subject to erosion, injection helps to prevent nutrient losses when soil particles are carried away by wind or by intensive rains. Therefore, fertilizers are usually injected after plowing or discing or when furrowing out when the soil is loose and plant residues are well dispersed.

Band placement — This method of application consists of placing fertilizer to the side of and / or below the seed or established plants. For dry materials a tube connected to the fertilizer hopper delivers the fertilizer to a furrow opened by a shoe or disc. For liquids the injection shank is usually attached to the seeder or to the cultivation equipment. The band applicator can be set to place the fertilizer in bands at any depth or position relative to the seed or plant. Depending on the plant, soil type and fertilizer, the band may be placed to the side and below the seed, or it may be placed directly below the seed. *Caution: Fertilizer placed too close to the seed may damage roots or inhibit seed germination. Placement with the seed is not recommended.*

Water-Run Application

Savings in time, labor, equipment and fuel costs are advantages of water-run fertilizer applications. Applications may be preplant or post-emergence, using either liquid or dry fertilizer materials. The plant nutrients should not be introduced into the system at the initiation of the irrigation set, as excessive leaching of certain nutrients may occur. Best results are obtained when the fertilizer application is gauged to enter the system toward the middle of the set and to terminate shortly before the set is completed. Application of fertilizers through the irrigation system in this manner prevents the nutrients from being leached beyond the reach of the roots or from lying near the surface where they may be inaccessible to plants. Precautions should be taken to avoid contamination of the water system and to minimize runoff when applying fertilizer in this manner.

Water-run applications include sprinkler, spitter, trickle, drip and dual-wall tubing systems. Sprinkler systems usually have all metal plumbing, while the others employ a considerable amount of plastic tubing and fittings.

Not all dry and liquid fertilizers can be applied through these systems. Filters are desirable when applying fertilizers through low-volume systems. They should be introduced at a point well ahead of the filters.

Dry nitrogen fertilizers can be applied provided (1) they do not contain coating materials that will most certainly plug the filter or system and (2) they are predissolved before being introduced into the system. Local fertilizer dealers should be consulted for information regarding coating materials on dry nitrogen fertilizers.

Since most dry phosphate fertilizers are not completely water-soluble, it is not recommended that they be applied through low-volume systems.

Dry potash fertilizers may be applied provided they are predissolved before being introduced into the system. Permanently placed spitters, spaghetti tubing in trickle systems, emitters in drip systems and surface or subsurface dual-wall systems are most suitable for potassium application. Permanently placed emission points permit a localized concentration of the potash fertilizer, providing downward movement into the soil by mass flow.

Liquid nitrogen fertilizers are very adaptable to all irrigation systems. In-line filters should be installed between the field storage tank and the injection pump. This will prevent most plugging problems.

There are several liquid phosphate fertilizers that can be used in these systems. Phosphorus has limited mobility in the soil. Therefore, the limitations indicated for potassium also apply to phosphorus. Factors other

Fig. 11-4. The homeowner in the upper photo is fertilizing his lawn by using a spinning-type spreader. The homeowner in the lower photo is fertilizing hers by using a hose-end attachment.

than mobility influence the probability of response with certain irrigation systems. For example, plant response to acidified phosphorus solutions applied through permanently placed, low-volume irrigation systems is possible because the nutrient is locally concentrated. Phosphorus moves downward through the soil by redissolution.

Liquid ammonium orthophosphate (e.g., 8-24-0) and ammonium polyphosphate (e.g., 10-34-0) fertilizers react with calcium in the water to produce insoluble calcium phosphate compounds that may plug the system. The higher the calcium content of the water, the faster the plugging rate. This problem can be eliminated by introducing an acid into the system at a rate that maintains the solution at about pH 5. Phosphoric acid can be used as a phosphate fertilizer in most systems without plugging problems by applying a rate which provides a slightly acid solution.

Special tanks are available on the market for dissolving dry fertilizers for water-run application. For liquid fertilizers, tanks can be set up at the irrigation pump site or injected directly into the irrigation line. Users of municipal water sources may wish to consult water authorities to determine if a check valve may be required to prevent fertilizer materials from being introduced into the municipal water system.

Foliar Application

A portion of a plant's nutritional needs may be met by directly applying a compatible fertilizer solution to the foliage. Foliar feeding is used in situations in which a quick response is required, such as when an unexpected deficiency develops during the growing season or where soil-applied nutrients were ineffective. Micronutrient deficiencies can frequently be corrected by foliar application. Primary and secondary nutrient needs can be partially met foliarly but should be supplemental to a sound soil-applied fertilizer program. Those interested in foliar feeding should seek the counsel of horticulturists, extension specialists or industrial agronomists regarding the plants to be treated, nutrients that can be applied and rates and methods of application.

Nutrients can be applied to plant foliage as sprays or through overhead sprinkler systems. Sprays are often preferred to the irrigation method because of more uniform distribution and less likelihood that the nutrients will be washed off before absorption occurs. In many situations, spray application is the only method that can be used. Equipment for applying pesticide sprays is usually used to also apply nutrients.

Ground spray equipment used for foliar feeding is usually of the high-pressure, low-gallonage type, designed to distribute the spray materials uniformly on the foliage and keep water volume to a minimum. The spray may be applied through single- or multiple-nozzle hand guns; through multiple-nozzle booms; by multiple-nozzle, oscillating or stationary cyclone-type orchard sprayers; or by backpack mist sprayers. Plant response may be affected by droplet size; therefore, the spray used must be regulated by adjusting pressure and selecting the proper nozzles and discs. The concentration of a nutrient in a foliar solution will vary widely with respect to the nutrient and plant in question. Low rates are generally used to prevent injury to the foliage. The addition of an adjuvant to the spray is recommended for better coverage.

If urea is used in foliar feeding, it should be of low biuret content. Biuret is a compound formed during the manufacture of urea. It is toxic to some plants and can be especially damaging when applied to foliage. Where sizable applications of urea are made, biuret should not exceed about 0.25 percent. Such solutions should also be buffered to below pH 7.0 to prevent ammonia burn.

Application of nutrients for foliar feeding through overhead irrigation systems follows some of the same principles observed for applying nutrients to the soil through irrigation. The exceptions are:

1. The concentration of the material being applied to the foliage must be considerably lower than when applied to the soil.
2. The nutrient must be introduced into the system toward the end of the irrigation set so that the nutrients remain on the foliage when the system is shut down.

Whether applied through ground equipment or overhead sprinkler systems, most plants respond better to foliar feeding when the nutrients are applied during the morning hours.

Somewhat related to foliar feeding is the application of nutrients to vines and fruit and nut trees during the latter part of the dormant season in combination with pesticide sprays. Methods of application of the nutrients are the same as those for foliar feeding. However, the nutrients may be applied at concentrations somewhat higher than those normally used.

CALIBRATION OF APPLICATION EQUIPMENT

The rate-of-delivery settings on fertilizer applicators may be predetermined by the equipment manufacturer. Charts showing the settings or orifice sizes for rates of application according to the kinds of fertilizer are usually affixed to the equipment or are given in the operator's manual. The local fertilizer supplier should be consulted for further information. Tables 11-1, 11-2 and 11-3 give selected calibration information for both liquid and dry fertilizers.

Table 11-1

Calibration of Fertilizer Application Machinery
(amounts of fertilizer per 100 feet of row)

Rate per:		Row Width				
Acre	1,000 Sq. Ft.	18 In.	24 In.	30 In.	36 In.	48 In.
---------- *(lbs.)* ----------		-------------------------------------- *(lbs.)* ---------				
250	5.7	⅞	1	1¼	1½	2
500	11.5	1¼	2	2½	3½	4½
750	17.2	2½	3	3¾	4½	7
1,000	23.0	3	4½	5¾	7	9
1,500	34.4	5	6½	8½	10½	14
2,000	45.9	6½	9½	11	13½	18
2,500	57.4	8½	11½	14½	17	23
3,000	68.9	10½	14	17½	21	28

Table 11-2

Calibration of Liquid Fertilizer Flow[1]

Gallons per Hour Wanted	Seconds to Fill 4-Ounce Jar	Seconds to Fill 8-Ounce Jar
½	225	450
1	112	224
2	56	112
3	38	76
4	28	56
5	22	44
6	18	36
7	16	32
8	14	28
9	12	24
10	11	22
12	9	18
14	8	16
16	7	14
18	6	12
20	5.5	11

[1] To calculate the rate of application required, the following formula can be used:

$$ft.^2 \times lbs. \text{ or gal. per ft.}^2 \div \text{set time} = lbs. \text{ or gal. per hour.}$$

Table 11-3

Calibration of Fertilizer Rate Through Sprinkler Irrigation Systems

Lateral Length in Feet	No. of Sprinklers (at 40-Foot Spacing)	Area Covered by 60-Foot Setting in Acres	Quantity to Apply per Setting, for Rate of 100 Pounds per Acre
160	4	0.22	22 lbs.
240	6	0.33	33 lbs.
320	8	0.44	44 lbs.
400	10	0.55	55 lbs.
480	12	0.66	66 lbs.
560	14	0.77	77 lbs.
640	16	0.88	88 lbs.
720	18	0.99	99 lbs.
800	20	1.10	110 lbs.
880	22	1.21	121 lbs.
960	24	1.32	132 lbs.

Example

To apply fertilizer at the rate of 300 lbs. per acre with 400 ft. of lateral moved at 60-ft. setting, lbs. of fertilizer applied at each setting of the lateral are calculated as follows:

Opposite a lateral length of 400 ft., find 55 lbs. of fertilizer to be applied per setting. Multiply 55 by 3 to give 165 lbs. to apply at each setting of the lateral for 300 lbs. per acre.

43,560 ft.^{2} = 1 acre.

FERTILIZER-PESTICIDE MIXTURES

Fertilizer-pesticide mixtures are used to fertilize plants and control soil-borne insects, diseases, nematodes and weeds. Fertilizer-pesticide mixtures are strictly regulated. Legislation, for example, provides that only qualified licensed individuals can formulate, sell, recommend and apply pesticides. Once a pesticide is added to a fertilizer, the mixture takes on a completely different set of legal ramifications regarding its use. It would be advisable for growers to consult with qualified and licensed technologists before mixing and / or applying fertilizer-pesticide combinations.

There are a number of fertilizer-pesticide products available through home and garden suppliers. Check label requirements for specific use instructions.

SUPPLEMENTARY READING

1. *Bulk Blend Quality Control Manual.* The Fertilizer Institute. June 1975.

2. *The Citrus Industry,* Volume III. W. Reuther. University of California. 1973.
3. *Fertilizer Technology and Usage.* M. H. McVickar, G. L. Bridger and L. B. Nelson. Soil Sci. Soc. of America. 1963.
4. *Nitrogen in Agricultural Soils.* F. J. Stevenson, ed. American Society of Agronomy. 1982.
5. *Potassium in Agriculture.* R. D. Munson, ed. American Society of Agronomy. 1985.
6. *The Role of Phosphorus in Agriculture.* F. E. Khasawneh, E. C. Sample and E. J. Kamprath, eds. American Society of Agronomy. 1980.
7. *Soil Fertility and Fertilizers,* Fourth Edition. S. L. Tisdale, W. L. Nelson and J. D. Beaton. The Macmillan Company. 1985.
8. TVA Fertilizer Bulk Blending Conference. Louisville, Ky. August 1 – 2, 1973.
9. *Using Commercial Fertilizers,* Fourth Edition. M. H. McVickar and W. M. Walker. The Interstate Printers & Publishers, Inc. 1978.

CHAPTER 12

Growing Plants in Solution Culture

The term *hydroponics* refers to the growing of plants with their roots immersed continuously or intermittently in a water solution containing the essential mineral elements. Quite commonly, glass wool, wood chips, pea gravel or coarse sand is used to give support to the plant roots while the solution is trickled through or flushed periodically to provide water and nutrients.

Scientists commonly use this method of growing plants in their research work. Food production for military forces was accomplished in this manner during World War II in remote areas where conditions were unfavorable for normal farming with soil. For many years there have been commercial operations that have utilized hydroponics to grow vegetables. With few exceptions, however, such commercial ventures have not proven economically viable.

SMALL-SCALE HYDROPONICS

Those who wish to try hydroponics on a small scale can do so with a fairly small investment in equipment. A 1- or 2-quart glass jar or crock may be used as the growing vessel. It is important that the container exclude light, or algae growth will be a problem. Glass or other transparent containers coated with black paint, aluminum foil or other opaque coating can be used to exclude light. Galvanized containers or those that rust should not be used unless they are lined with polyethylene or similar plastic. A plastic food bag works well for this purpose.

Fill the container with the nutrient solution to within 1½ inches from the top. Fit the top of the container with a large cork with a ¾-inch hole in the center. Suspend the seedling or cutting with its roots in the solution. Loosely pack non-absorbent cotton or glass wool around the stem to support the plant and to allow air to circulate through the hole. It may be advantageous to use a small aquarium pump to keep the solution aerated. Many plants can be grown together if a larger container is used.

For large-scale water culture, use a wooden, concrete, glass or other suitable tank at least 6 inches deep and 2 to 3 feet wide. Polyethylene or nontoxic asphalt paint can be used to line the tank. Use a board or other adaptable material to support the plants over the nutrient solution. Another method is to use wire mesh covered with a mat of moss, excelsior, or similar material. Use of an air pump is again desirable.

As the nutrient solution level drops due to plant extraction and evaporation, add solution to keep the roots submerged. The solution pH should be monitored and maintained in the 5 to 7 range. Every two to three weeks, discard the old solution and replace it with fresh solution.

An alternative system is to rear plants in clean, acid-washed sand in a flower pot or other suitable container with a drainage hole. The pots are flooded daily with nutrient solution and the drainage collected in a reservoir. This is then reapplied daily or as needed, and, after two or three weeks, discarded. Just before renewing the solution, irrigate plants with fresh water to flush out any accumulated salts.

A variation of this method employs a small pump and timer set to periodically recycle the solution back into the pots. In another, a container is supported above the pots and the solution is applied slowly through drip tubes. As above, the drainage is recycled through the system.

Most water in the home is satisfactory unless it is treated with a water softener, which adds too much sodium to the system. It is much better to use hard water, which contains calcium and magnesium. Distilled or deionized water can be used, but micronutrients will have to be added. Ordinary water often contains some zinc, copper and other nutrients.

FORMULAS FOR SMALL-SCALE SYSTEMS

The formula shown in Table 12-1 was developed by Professor Hoagland at the University of California. The chemicals can be obtained from chemical supply houses, drug stores and garden supply dealers. Technical and fertilizer grade salts are adequate unless nutrient deficiency research is being conducted. In this case, the experimenter will need chemically pure salts, distilled or deionized water and plastic lined pots.

Dissolve the potassium phosphate first and then add the other chemicals in the order given. Dissolve each one before adding the next. Warm water will speed up the process.

Table 12-1

Components of the Hoagland Solution

Salt	Nutrient(s)	Amount for 25 Gallons		
		(g)	(oz.)	(tbsp.)
Potassium phosphate, KH_2PO_4 (monobasic salt)	P and K	14	½	1
Potassium nitrate, KNO_3 (fertilizer grade)	K and N	57	2	4
Calcium nitrate, $5Ca(NO_3)_2 \cdot NH_4NO_3 \cdot 10H_2O$ (fertilizer grade)	N and Ca	85	3	7
Magnesium sulfate, $MgSO_4 \cdot 7H_2O$ (Epsom salts)	Mg and S	43	1½	4

Essential micronutrients (also called *trace* or *minor elements*) must be added. It is best to make stock solutions of these and then add the suggested amount to the Hoagland nutrient solution. This micronutrient solution is made up as follows:

ADD TO 1.5 GALLONS OF WATER:
Powdered Boric Acid . . . . .3 Tsp
Manganese Chloride or Manganese Sulfate . . . . .1 Tsp

Add ½ pint of this stock solution to 25 gallons of complete nutrient solution. There will probably be enough impurities in the various salts and water to provide the needed zinc and copper.

The iron stock solution is also prepared separately by dissolving 1½ ounces of iron chelate in 1 gallon of water. Add ½ pint of this iron solution to 25 gallons of complete nutrient solution once or twice a week during the growing period. Alternate sources of iron are iron tartrate, iron citrate and ferrous sulfate. Use 4 level teaspoons of one of these sources per gallon of water and use as directed for the chelated iron. Remember to keep the stock solutions in the dark or in opaque containers to prevent troublesome algae growth.

LARGE-SCALE HYDROPONICS

For commercial-type enterprises, hydroponics becomes more complex and requires close attention. It is not an inexpensive way to grow plants, although this will vary with the sophistication of the equipment, the size of the operation and the experience of the operator.

Commercial enterprises commonly prefer the "closed" system of hydroponics (see Figure 12-1). The nutrient solution is periodically pumped from holding containers up through pea gravel beds in which the plants are rooted. Gravity drainage returns the solution to the tank, where it remains until the next pumping cycle. The frequency of pumping is related to the size of the plants, temperature, relative humidity and drainage rate.

Because of the large investment in commercial hydroponics systems, it is wise to monitor the nutrient concentration and supplement it as needed. As previously stated, solution pH should also be monitored and maintained between 5 and 7. Usually the solution is renewed after two or three weeks, but extended use of the solution is possible if a careful monitoring system is established and proper nutrient salt additions are made.

Sanitation and disease monitoring are vital since the hydroponic solutions are usually common to all the plants in the system. Once an infestation occurs, the solution can rapidly spread a disease. Also, if the solution is not changed regularly there is a good chance of specific salt buildup. This may come from the pipes, from the water source or from impurities in the chemicals used.

Fig. 12-1. Tomatoes growing in a commercial hydroponic greenhouse.

Commercial preparations of premixed salts are available for hydroponics, but no single formula will fit all plants and all growing conditions. The operator who desires to extend the use of the solutions will need a supply of individual salts to add as laboratory analysis dictates.

FORMULAS FOR LARGE-SCALE SYSTEMS

Liquid stock solutions, I and II, are prepared separately since certain of the components will react forming precipitates when mixed together in concentrated form.

STOCK SOLUTION I (Table 12-2) contains nitrogen, phosphate and potassium plus magnesium, sulfur and the micronutrients. Preparation of Stock Solution I is a multiple-step process. Predissolve the potassium and magnesium salts in ¾ the desired amount of water. The micronutrient concentrate will be added as the final step.

To make the MICRONUTRIENT CONCENTRATE (Table 12-3), in a separate vessel add boric acid to ⅓ the volume of water. Boil until dissolved, and cool. Next, dissolve the other micronutrient salts in another container with about ⅔ of the volume of water. Combine with the boric acid solution and bring to final volume.

Preparation of Stock Solution I is then completed by adding the stated quantity of micronutrient concentrate to the potassium and magnesium salt solution (Table 12-2).

STOCK SOLUTION II (Table 12-4) is prepared by thoroughly mixing the iron chelate (Sequestrene 330 Fe) in a little water, adding it to the dissolved calcium nitrate and bringing to volume.

The complete nutrient solution is then prepared by separately pre-diluting stock solutions I and II by 1:200 with water and combining. Do not mix the concentrated stock solution without pre-dilution, since calcium phosphate will precipitate. Table 12-5 shows the amount of each stock solution required for three final, complete nutrient solution volumes.

The concentration of nutrient elements in the final hydroponic solution based upon the 1:200 dilution is as follows:

	N	P	K	Ca	Mg	S	Fe	B	Mn	Zn	Cu	Mo
ppm	119	30	140	100	24	32	2.5	0.25	0.25	0.025	0.01	0.005
me / L	8.5	1	3.5	5	2	2						

It is helpful to know the final diluted concentration, since laboratories often report concentrations in parts per million (ppm) or milliequivalents

Table 12-2

A Commonly Used Primary Nutrient Stock Solution
(Stock Solution I)

Salt	For 50 Gallons	For 10 Liters
Potassium nitrate (KNO_3)	21 lbs.	503 g
Potassium phosphate (KH_2PO_4)	12 lbs.	288 g
Magnesium sulfate ($MgSO_4 \cdot 7H_2O$)	21 lbs.	503 g
Micronutrient concentrate	5 gal.	1,000 ml

Table 12-3

A Commonly Used Micronutrient Concentrate

Salt	For 50 Gallons	For 10 Liters
Boric acid (H_3BO_3)	54g	2.8g
Manganese sulfate ($MnSO_4 \cdot H_2O$)	28g	1.5g
Zinc sulfate ($ZnSO_4 \cdot 7H_2O$)	4g	0.2g
Copper sulfate ($CuSO_4 \cdot 5H_2O$)	1g	0.05g
Molybdic acid ($H_2MoO_4 \cdot H_2O$)	0.5g	0.03g

Table 12-4

A Commonly Used Stock Solution Containing Nitrogen,
Calcium and Iron (Stock Solution II)

Salt	For 50 Gallons	For 10 Liters
	(lbs.)	*(g)*
Calcium nitrate[1]	45	1,079
Sequestrene 330 Fe[2]	2	48

[1] Commercial fertilizer grade.

[2] A Ciba-Geigy Chemical Company product. Other iron chelates may work equally well but should be tested before they are used.

Table 12-5

Amount of Stock Solutions Needed for Different Volumes

Stock Solution[1]	Final Volume		
	20 Liters	100 Gallons	1,000 Gallons
I (See Table 12-2)	100 ml	2 qt.	5 gal.
II (See Table 12-4)	100 ml	2 qt.	5 gal.

[1] *IMPORTANT:* Pre-dilute 1:200 with water.

Table 12-6

Suggested Salts for Formula Modifications and Amounts to Add

Salt	Amount Required to Add to 50 Gallons Stock Solution to Give 1 me / L When Diluted 200-fold	
	(nutrient)	(lbs.)
Potassium phosphate (KH_2PO_4)	P	11.4
	K	11.4
Diammonium phosphate [$(NH_4)_2HPO_4$]	P	11.0
	N	5.5
Phosphoric acid, fertilizer grade (H_3PO_4—52% P_2O_5)	P	11.4
Phosphoric acid, reagent grade (85% H_3PO_4)	P	9.6
Calcium nitrate, fertilizer grade	Ca	9.0
[$5Ca(NO_3)_2 \cdot NH_4NO_3 \cdot 10H_2O$][1]	N	7.5
Calcium nitrate, reagent grade [$Ca(NO_3)_2 \cdot 4H_2O$][1]	Ca	9.8
	N	9.8
Ammonium nitrate (NH_4NO_3)	N	3.3
Potassium nitrate (KNO_3)	N	8.4
	K	8.4
Nitric acid (70% HNO_3)	N	7.5
Urea [$CO(NH_2)_2$]	N	2.5

[1] REMEMBER: Do not add calcium nitrate to concentrated solutions containing phosphates.

per liter (me / L). This will also provide a base from which to calculate how much nutrient need be added to replenish nutrients in the solution, or to substitute salts. For example, if the water used in the system contains 100 ppm calcium, then calcium nitrate can be left out and a substitute salt such as ammonium nitrate used. Since calcium nitrate supplies 5 me of calcium and nitrogen, then 5 me of nitrogen is needed to supply that which would come from calcium nitrate. Table 12-6 shows the amount of alternate salts required to add to 50 gallons of stock solution to give 1 me / L when diluted 200-fold. Note that this requires 3.34 pounds of ammonium nitrate per me. Thus for 5 me it will take 16.7 pounds of ammonium nitrate in Stock Solution II to substitute for the calcium nitrate.

The material covered in this chapter is not intended as a complete list, nor does it include all crops and systems. Observance of the guidelines listed, however, will serve as an aid in avoiding common problems.

MONITORING NUTRIENT SOLUTIONS

The nutrient formulas can be quite well monitored by the use of a salt bridge or conductivity meter. Since this instrument measures the electrical conductivity (EC) of the solution, which in turn is a function of the nutrients in solution, it provides a measure of the decline in mineral nutrients as they are used by the plants. A base point should be established by measuring the solution conductivity immediately after the salts are added and mixed thoroughly. Additional measurements should be made every two or three days and more frequently as the plants increase in size. The EC indicates the concentration of the total ions in solution, not individual ions. Experience has shown, however, that there is a good correlation between EC and nitrate nitrogen, phosphate-phosphorus and potassium. As the EC decreases, fertilizer salt or stock solution should be added to bring the EC back to the base level. Bring the volume of solution to nearly the original level with water before taking the EC reading, particularly with larger, faster-growing plants.

The quantity of stock solution to add is related to the growth rate of the plants. Usually about 1 gallon of stock solution needs to be added per day per 1,000 gallons of solution volume for plants 6 to 8 feet tall. During the first month, while plants are small, the requirements are one-fourth or less. Adjustments are made based upon the EC reading and experience. Make sure that the solution is thoroughly mixed before EC measurements are made.

Water quality is an important factor to consider in hydroponic culture. Where the water has a high salt content, it may limit the time it can be used. Some water may be unusable because it contributes excess or toxic salts. To avoid problems it is a good practice to have a water quality test made by a reliable laboratory.

Solution volume should be kept to 4 or 5 gallons per plant for tomatoes. This may be somewhat less for other plants, but care should be taken to provide adequate solution. Quick changes in solution concentration are thus avoided, and the margin of safety in culture is extended.

Tissue analysis is another monitoring system that should be performed periodically. Again, a reliable laboratory should be used; general guidelines for sampling are described in Chapter 10 along with critical levels for various plants.

Consultation with the proper authorities of the Cooperative Extension Service or university will be of assistance. References concerning this method of growing plants are listed below as Supplementary Reading.

SUPPLEMENTARY READING

1. "Growing Plants in Nutrient Cultures." J. G. Seeley. *Horticulture Science.* August 1974.
2. *Growing Plants in Solution Culture.* E. Epstein and B. A. Krantz. AXT-196. University of California. 1972.
3. *Guide Values for Nutrient Element Contents of Vegetables and Flowers Under Glass.* C. De Kreij, C. Sonneveld and M. G. Warmenhoven. Glasshouse Research and Experiment Station, Naaldwijk, The Netherlands. No. 15, March 1987.
4. *Hydroponics: A Guide to Soilless Culture Systems.* H. Johnson, Jr. Leaflet 2947. University of California. 1977.
5. *Hydroponics as a Hobby — Growing Plans Without Soil.* J. D. Butler and N. F. Oebker. Circular 844. University of Illinois. 1974.
6. *A Method for Calculating the Composition of Nutrient Solutions for Soilless Cultures.* C. Sonneveld. No. 10, Second Translated Edition, Glasshouse Crops Research and Experiment Station, Naaldwijk, The Netherlands, 1985.
7. *Soilless Growth of Plants,* Second Edition. C. Ellis and M. W. Swaney. Van Nostrand Reinhold Company. 1974.

CHAPTER 13

Benefits of Fertilizers to the Environment

Research has shown that good fertilizer practices that match fertilizer inputs to plant requirements will achieve high yields and quality and also be beneficial to the environment. However, fertilizer applied improperly or in excessive amounts can lead to potential environmental problems.

The concern for the environment has led to the creation of numerous agencies and bureaus whose roles directly or indirectly have a bearing on fertilizers and their manufacture, storage, handling and transportation. The authority of these organizations is subject to change. Names and addresses are given, and any company or individual working closely in the areas affected by these organizations should write for current detailed information.

HOW THE PROPER USE OF FERTILIZER IMPROVES THE ENVIRONMENT

Fertilizer improves the environment in several ways. It:

1. Provides cleaner air.
2. Cuts down on soil erosion, leading to cleaner surface waters.
3. Reduces pollution of ground waters.
4. Leaves more land for open spaces and recreational purposes.
5. Beautifies.

Provides Cleaner Air

Growing plants, through a process called *photosynthesis,* convert carbon dioxide into carbohydrates and at the same time give off oxygen. These plants take in the carbon dioxide through stomata, which are microscopic openings in their leaves, and give off life-supporting oxygen. People, animals and industry consume large quantities of oxygen and generate large amounts of carbon dioxide. The plants on the earth and in the oceans maintain the oxygen:carbon dioxide ratio in the air. Air contains about 20 percent oxygen and 0.03 percent carbon dioxide. This balance is a delicate one. Larger plants use more carbon dioxide and give off more oxygen (Figure 13-1). Proper fertilization stimulates plant growth and thus indirectly increases the amount of oxygen given off.

Still more important to people, however, is the fact that productive soils remove carbon monoxide from the air. Scientists have identified the *Aspergillus* and *Penicillium* types of fungi as being responsible for carbon monoxide removal. These organisms, most abundant in fertile soils, are helping to provide cleaner air by using for their metabolism large amounts of the carbon monoxide created annually by people.

Gas chromatographic studies also show that fertile soils have the capacity to absorb substantial quantities of sulfur dioxide and hydrogen sulfide. These experiments involving both steam-sterilized and non-steam-sterilized soils indicate that soil microorganisms play little, if any, part in the absorption of sulfur dioxide and hydrogen sulfide. The research reveals that moist, fertile soils can absorb from 2.1 to 15.2 pounds of sulfur dioxide per 1,000 square feet. In brief, these investigations indicate that soil is an important natural sink for gaseous atmospheric pollutants and may

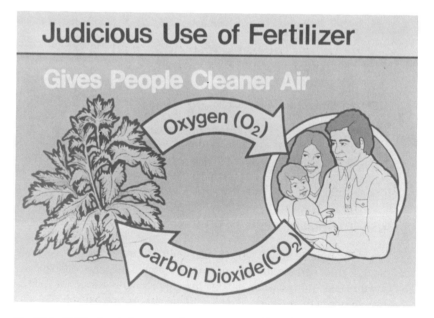

Fig. 13-1. Well-fertilized plants provide more oxygen and consume more carbon dioxide.

prove valuable for purification of industrial emissions heavily polluted by sulfur gases.

Cuts Down on Soil Erosion

Well-fertilized plants have both extensive tops and roots. A well-developed and extensive top growth reduces the pounding effect of natural raindrops or sprinklers. Instead of breaking down the structure of the soil surface, the water trickles down into the soil (Figure 13-2). Further, well-

Fig. 13-2. Soil erosion can become a problem when land is left unprotected.

established plantings with extensive root systems hold the soil in place, reducing runoff and minimizing erosion (Figure 13-3). With minimal soil erosion, streams run clear and clean.

Research has shown that soil erosion is the major way phosphates are lost from the soil. The phosphorus is carried off as adsorbed and precipitated phosphate. Soil sediment also contains nitrogen and other nutrients, so any reduction in erosion also reduces the levels of these nutrients in surface water.

Fig. 13-3 Roots of well-fertilized plants hold soil in place.

Reduces Pollution of Ground Waters

Nitrates, along with other soluble nutrients such as chlorides and sulfates, are associated with underground water pollution. As previously mentioned, phosphates can normally be eliminated from discussion since even completely water-soluble forms, when applied to the soil, are rapidly converted to water-insoluble forms and therefore do not move with the soil solution. Phosphate stays essentially where it is placed in the soil, and there is little or no leaching. This is why fertilizer-phosphate placement within the root zone is a commonly recommended practice. If erosion is controlled, phosphates stay where placed.

Regardless of the form of nitrogen applied to the soil, essentially all of it will eventually be converted to the nitrate form. Nitrate nitrogen moves downward, laterally or upward, depending on the movement of soil moisture. For example, if nitrates in the soil solution move below the root zone, they may eventually enter the ground water. There are natural processes such as denitrification that may reduce or eliminate such ground water contamination. But active management of irrigation and drainage water by landscapers, nursery workers, groundskeepers, etc., will significantly improve efficiency and reduce water and nutrient loss.

Plants with extensive root systems — the kind that develop when well fertilized — utilize the soluble nutrients including nitrates as the soil solution moves either downward or upward within the root zone of the plants. The deeper and more extensive the root system, the more effective is the planting in keeping soluble nutrients from reaching the ground waters.

An active and strong root system acts like a wick. It not only takes up water in the immediate area but also, through upward capillary action, draws water along with soluble materials from deep in the soil (Figure 13-4). This action helps reduce the potential for pollution of ground waters.

Fig. 13-4. Active root systems intercept nutrients before they reach groundwater.

Leaves More Land for Open Spaces and Recreational Purposes

Fertilizer is responsible for approximately 35 percent of all agricultural and horticultural production. Without the use of fertilizer, growers would have to devote much more land area to production. This would mean not only less open land, but also increased soil erosion.

Several studies in the United States have been conducted to determine the degree to which fertilizer substitutes for land. Donald Ibach of the USDA estimated that 1 ton of nutrients applied annually would substitute for 9.4 acres of land. The degree of substitution varies, of course, depending on the rate of fertilizer application, plant response, types of management practices and other variables.

Recreational areas are very important to much of the population in and around large urban centers. Well-kept parks, playgrounds, golf courses, street plantings and home gardens are important factors in responsibly developed urban and suburban areas (Figure 13-5).

Good turf is a basic ingredient for recreational and aesthetic needs. Healthy turf also protects environmental quality. Well planned and properly managed landscaping prevents dust from blowing, stops water erosion, filters sediment from storm runoff and keeps streams, walkways and roads cleaner (Figure 13-6).

Trees and shrubs can provide much-needed shade, reduce noise and enhance the beauty and value of public and private lands. These plantings depend on good growth for their value. Such growth requires careful planting and maintenance, including the proper use of fertilizers.

Food, water and air are among the most basic needs. But our protection of the quality of the soil, water and air provides more opportunity for each person to enjoy a higher quality of life.

Fertilizer is the most profitable input of all those which growers use in their production program. Thus, fertilizer is basic to a quality product for

Fig. 13-5. Fertilizer is essential for maintaining quality parks, golf courses, etc.

Fig. 13-6. Through the use of fertilizer, more land can be used for recreation areas and wildlife preserves.

Fig. 13-7. Fertilizers are credited with over a third of our food production.

everyone at an affordable price. It lowers the unit cost of production, which lowers the cost to the consumer. With this basic need met, we can look forward to a better quality of life.

MATCHING FERTILIZER INPUTS TO PLANT NEEDS

Proper fertilization means matching fertilizer inputs to plant and soil needs. It calls for:

1. Using the *right nutrients.*
2. Using the *right amount* of the nutrients.
3. Applying the nutrients in the *right place.*
4. Applying the nutrients at the *right time.*

There are several "tools" available to help the grower match fertilizer inputs to plant and soil needs. These include soil and tissue testing, discussed in Chapter 10, and plant food requirements of various crops, discussed in Chapter 4. Past production records plus management skills are important to the grower in deciding which fertilizers and how much to use to achieve optimum quality and production without using excess.

PROFESSIONAL ORGANIZATIONS

Professional groups in support of the horticulture and landscaping industry include:

American Association of Nurserymen
1250 I Street, N.W. — Suite 500
Washington, DC 20005

American Council for Turfgrass
Soil & Crop Sciences Center
Texas A&M University
College Station, TX 77843

American Horticultural Society
P.O. Box 0105
Mt. Vernon, VA 22121

American Society for Horticultural Sciences
701 N Street — Asaph Street
Alexandria, VA 22314

American Society of Landscape Architects
1733 Connecticut Avenue, N.W.
Washington, DC 20009

California Association of Nurserymen
1419 21st Street
Sacramento, CA 95814

California Fertilizer Association
1700 I Street — Suite 130
Sacramento, CA 95814

California Landscape Contractors Association
2226 K Street
Sacramento, CA 95816

Golf Course Superintendents' Association of America
1617 St. Andrews Drive
Lawrence, KS 66044

Hobby Greenhouse Association
5439 Palomino Court
Humble, TX 77338

Horticultural Research Institute
1250 I Street, N.W. — Suite 500
Washington, DC 20005

AGENCIES AND BUREAUS VESTED WITH FERTILIZER CONTROLS AND REGULATIONS

The scope and activities of fertilizer control with respect to meeting fertilizer guarantees are covered in Appendix B. As previously mentioned, recent years have seen the creation of several new agencies, bureaus and organizations, most of which deal in one way or another with the quality of our environment. These organizations can have an effect on the manufacture, storage, handling, transportation and use of fertilizer. Current detailed information can be obtained by writing directly to these organizations listed below and to other local and state agencies in your area.

National Office
Occupational Safety and Health Administration (OSHA)
U.S. Department of Labor
200 Constitution Avenue, N.W.
Washington, DC 20210

Regional Office
Occupational Safety and Health Administration (OSHA)
11349 Federal Building
450 Golden Gate Avenue
P.O. Box 36017
San Francisco, CA 94102

Environmental Protection Agency (EPA)
401 M Street, S.W.
Washington, DC 20460

Environmental Protection Agency (EPA) Regional Offices:
Region 5 — Minn., Wis., Mich., Ill., Ind., Ohio
230 South Dearborn Street
Chicago, IL 60604

Region 6 — Tex., Ark., La., N. Mex., Okla.
1445 Ross Avenue — 12th Floor
Dallas, TX 75202-2733

Region 7 — Iowa, Kans., Mo., Nebr.
726 Minnesota Avenue
Kansas City, MO 66101

Region 8 — Colo., Mont., N. Dak., S. Dak., Utah, Wyo.
One Denver Place
999 18th Street — Suite 1300
Denver, CO 80202-2413

Region 9 — Hawaii, Nev., Ariz., Calif., Guam, Am. Samoa
215 Fremont Street
San Francisco, CA 94105

Region 10 — Wash., Oreg., Ida., Alas.
1200 Sixth Avenue
Seattle, WA 98101

U.S. Department of Transportation (DOT)
Office of Hazardous Materials
400 7th Street, S.W.
Washington, DC 20590

Chemical Transportation Center (CHEMTREC)
1825 Constitution Avenue, N.W.
Washington, DC 20009

SUPPLEMENTARY READING

1. *Contributions of Fertilizer to the Economy and the Environment.* L. B. Nelson. TVA Publication. 1972.

2. *Facts from Our Environment.* Potash & Phosphate Institute. Special Bulletin. 1972.
3. "Nitrate Breakdown." *Agricultural Research.* May 1970.
4. "Nitrogen Facts and Fallacies." W. L. Garman. *Plant Food Review,* No. 1. 1969.
5. *Nitrogen Management and the Environment.* Potash & Phosphate Institute. 1988.
6. "Sorption of Gaseous Atmospheric Pollutants by Soils." *Soil Sci.* Vo. 116, No. 4: 313–319. 1973.
7. *Urban Forestry, Planning and Managing Urban Vegetation.* Robert W. Miller. Prentice Hall. 1988.
8. *The Vital Role of Phosphorus in Our Environment.* Potash & Phosphate Institute. 1987.

APPENDIX A

Glossary of Terms

AAPFCO — American Association of Plant Food Control Officials.

ABSORPTION — The process by which a substance is taken into and included within another substance, e.g., intake of water by soil, or intake of gases, water, nutrients or other substances by plants.

ACID-FORMING — A term applied to any fertilizer that tends to make the soil more acid.

ACID SOIL — A soil with a pH value below 7.0. A soil having a preponderance of hydrogen over hydroxyl ions in the soil solution.

ACTIVATED SEWAGE SLUDGE — An organic fertilizer made from sewage freed from grit and coarse solids and aerated after being inoculated with microorganisms. The resulting flocculated organic matter is withdrawn from the tanks, filtered with or without the aid of coagulants, dried, ground and screened.

ADSORPTION — The increased concentration of molecules or ions at a surface, including exchangeable cations and anions on soil particles.

AERATION, SOIL — The exchange of air in soil with air from the atmosphere. The composition of the air in a well-aerated soil is similar to that in the atmosphere; in a poorly aerated soil, the air in the soil is considerably higher in carbon dioxide and lower in oxygen than the atmosphere above the soil.

AGGREGATE — A group of soil particles cohering so as to behave mechanically as a unit.

ALKALINE — A basic reaction in which the pH reading is above 7.0, as distinguished from acidic reaction, in which the pH reading is below 7.0.

ALKALINE SOIL / MEDIA — A soil / media that has an alkaline reaction, i.e., a soil for which the pH reading of the saturated soil paste is above 7.0.

ALKALI SOIL / MEDIA — See SODIC SOIL / MEDIA.

AMENDMENT — Any material, such as lime, gypsum, sawdust or synthetic conditioners, that is worked into the soil to make it more productive. Strictly, a fertilizer is also an amendment, but the term *amendment* is used more commonly for added materials other than fertilizer.

AMINO ACIDS — Nitrogen-containing organic compounds, large numbers of which link together in the formation of the protein molecule. Each amino acid molecule contains one or more amino ($-NH_2$) groups and at least one carboxyl ($-COOH$) group. In addition, some amino acids (cystine and methionine) contain sulfur.

AMMONIATED SUPERPHOSPHATE — A product formed by ammoniating superphosphate.

AMMONIATION — A process wherein ammonia (anhydrous, aqua or a solution containing ammonia and other forms of nitrogen) is used to treat superphosphate to form ammoniated superphosphate, or to treat a mixture of fertilizer ingredients (including phosphoric acid) in the manufacture of a multinutrient fertilizer.

AMMONIFICATION — Formation of ammonium compounds or ammonia.

AMMONIUM CITRATE [$(NH_4)_3C_6H_5O_7$] — A salt formed from ammonia and citric acid. A neutral ammonium citrate solution, prepared by the official methods of the AOAC, is used as a reagent in the determination of "available" phosphoric acid in fertilizers. After a sample is washed with water to remove the water-soluble phosphoric acid (P_2O_5), the residue is treated with the neutral ammonium citrate solutions, as prescribed by the official methods, and the phosphoric acid removed by this extraction is termed "citrate-soluble." The sum of the water-soluble plus the citrate-soluble phosphoric acid is termed "available."

ANALYSIS — The percentage composition as found by chemical analysis, expressed in those terms that the law requires and permits. Although "analysis" and "grade" sometimes are used synonymously, the term "grade" is applied only to the three primary plant foods — nitrogen (N), available phosphate (P_2O_5) and potash (K_2O) — and is stated as the guaranteed minimum quantities present. (See also GRADE.)

ANGLE OF REPOSE — The angle between the horizontal and the slope of a pile of loose material at equilibrium.

ANION — An ion carrying a negative charge of electricity.

ANNUAL — Horticulturally, a plant that completes its entire life cycle in a single growing season.

AOAC — Association of Official Analytical Chemists (of North America).

APATITE (rock phosphate) — A mineral phosphate having the type formula $Ca_{10}(X_2)(PO_4)_6$ where X is usually fluorine, chlorine or the hydroxyl group, either singly or together. Fluorapatite is widely distributed as the crystalline mineral and as amorphous phosphate rock, both forms of which are important fertilizer materials. Crystalline fluorapatite contains from 38.0 to 41.0 percent phosphoric acid (P_2O_5) and from 3.2 to 4.3 percent fluorine. Calcium hydroxyapatite or calcium hydroxy-phosphate, $Ca_{10}(OH)_2(PO_4)_6$, may be formed to a small extent in ammoniated superphosphate.

ARBORICULTURE — Cultivation of woody plants, particularly those used for decoration and shade.

ARTIFICIAL MEDIA — A mixture of various organic and inorganic constituents, such as perlite, vermiculite and peat moss, but not including soil, which is used for growing plants in containers or beds.

AVAILABLE — In general, a form capable of being assimilated by a growing plant. Available nitrogen is defined as the nitrogen that is water-soluble plus what can be made soluble or converted into free ammonia. Available phosphoric acid is that portion which is water-soluble plus the part which is soluble in ammonium citrate.

Available potash is defined as that portion soluble in water or in a solution of ammonium oxalate.

AVAILABLE NUTRIENT IN SOIL / MEDIA — The part of the supply of a plant nutrient in the soil that can be taken up by plants at rates and in amounts significant to plant growth.

AVAILABLE WATER IN SOIL / MEDIA — The part of the water in the soil that can be taken up by plants at rates significant to their growth; usable; obtainable.

BASE EXCHANGE — The replacement of cations, held on the soil complex, by other cations. (See also CATION EXCHANGE CAPACITY.)

BASIC SLAG — A by-product in the manufacture of steel, containing lime, phosphate and small amounts of other plant food elements such as sulfur, manganese and iron. Basic slags may contain from 10 to 17 percent phosphate (P_2O_5), 35 to 50 percent calcium oxide (CaO) and 2 to 10 percent magnesium oxide (MgO). The available phosphate content of most American slag is in the range of 8 to 10 percent.

BASIC SOIL / MEDIA — See ALKALINE SOIL / MEDIA.

BONE MEAL — Raw bone meal is cooked bones ground to a meal without any of the gelatin or glue removed. Steamed bone meal has been steamed under pressure to dissolve out part of the gelatin.

BRAND — The trade name assigned by a manufacturer to a particular fertilizer product.

BRIMSTONE — Sulfur.

BUFFER CAPACITY OF SOIL — The ability of the soil to resist a change in its pH (hydrogen ion concentration) when acid-forming or base-forming materials are added to the soil.

BULK BLENDING — The practice of mixing dry, individual, granular materials or granulated bases. The product is a mixture of granular materials rather than a granulated mixture.

BULK DENSITY — The ratio of the mass of water-free soil to its bulk volume. Bulk density is expressed in pounds per cubic foot or grams per cubic centimeter and is sometimes referred to as *apparent density.* When expressed in grams per cubic centimeter, bulk density is numerically equal to apparent specific gravity or volume weight.

CALCAREOUS SOIL — A soil containing calcium carbonate, or a soil alkaline in reaction because of the presence of calcium carbonate; a soil containing enough calcium carbonate to effervesce (fizz) when treated with dilute hydrochloric acid.

CALCIUM CARBONATE EQUIVALENT — The amount of calcium carbonate required to neutralize the acidity produced by a given quantity of fertilizer product.

CARBOHYDRATE — A compound containing carbon, hydrogen and oxygen. Usually the hydrogen and oxygen occur in the proportion of 2 to 1, such as in glucose ($C_6H_{12}O_6$).

CARBON:NITROGEN RATIO — The ratio obtained by dividing the percentage of organic carbon by percentage of nitrogen.

CATION — An ion carrying a positive charge of electricity. Common soil cations are calcium, magnesium, sodium, potassium and hydrogen.

CATION EXCHANGE CAPACITY — The total quantity of cations which a soil can adsorb by cation exchange, usually expressed as milliequivalents per 100 grams. Measured values of cation exchange capacity depend somewhat on the method used for the determination.

CHELATES — Certain organic chemicals, known as *chelating agents*, form ring compounds in which a polyvalent metal is held between two or more atoms. Such rings are chelates. Among the best chelating agents known are ethylenediaminetetraacetic acid (EDTA), hydroxyethylenediaminetriacetic acid (HEDTA) and diethylenetriaminepentaacetic acid (DTPA). Citric acid is also used as a chelating agent.

CHLOROSIS — Yellowing of green portions of a plant, particularly the leaves.

CITRATE-SOLUBLE PHOSPHORIC ACID — That fraction of the phosphoric acid insoluble in water but soluble in neutral ammonium citrate. However, since that soluble in water is also soluble in ammonium citrate, "citrate-soluble" may be used to indicate the sum of water-soluble plus citrate-soluble phosphoric acid. (See also AVAILABLE.)

CLAY — A minute soil particle less than 0.002 millimeter in diameter.

COATED FERTILIZERS — Fertilizer materials, generally urea, that are coated to slow the release of the fertilizer. Coating material is most commonly sulfur, but resins and thermoplastics are also used.

COLLOID — The soil particles (inorganic or organic) having small diameters ranging from 0.20 to 0.005 micron. Colloids are characterized by high base exchange.

COMPLETE FERTILIZER — A fertilizer containing all three of the primary fertilizer nutrients (nitrogen, phosphate and potash) in sufficient amounts to be of value as nutrients.

COMPOST — A mixture that consists largely of decayed organic matter and is used for fertilizing and conditioning soil.

CONDITIONER (of fertilizer) — A material added to a fertilizer to prevent caking and to keep it free-flowing.

CONDUCTIVITY, ELECTRICAL — A physical quantity that measures the readiness with which a medium transmits electricity. Commonly used for expressing the salinity of irrigation waters and soil extracts because it can be directly related to salt concentration. It is expressed in decisiemens per meter (dS / m), or in millisiemens per centimeter (mS / cm) or millimhos per centimeter (mmhos / cm), at 25°C.

CONTAINER STOCK — Nursery plants grown entirely in containers rather than being dug from a field.

CURING — The process by which superphosphate or mixed fertilizers are stored until the chemical reactions have run to, or nearly to, completion.

CYTOPLASM — The portion of the protoplasm of a cell outside the nucleus.

DAMPING-OFF — Sudden wilting and death of seedling plants resulting from attack by microorganisms.

DENITRIFICATION — The process by which nitrates or nitrites in the soil or organic deposits are reduced to lower oxides of nitrogen by bacterial action. The process results in the escape of nitrogen into the air.

DOLOMITE — A material used for liming soils in areas where magnesium and calcium are needed. Made by grinding dolomitic limestone, which contains both magnesium carbonate, $MgCO_3$, and calcium carbonate, $CaCO_3$. (See also LIME.)

DRIP IRRIGATION — See LOW-VOLUME IRRIGATION.

ECOLOGY — The branch of biology that deals with the mutual relations among organisms and between organisms and their environment.

ELEMENTAL GUARANTEES — See GUARANTEES.

ENVIRONMENT — All external conditions that may act upon an organism or soil to influence its development, including sunlight, temperature, moisture and other organisms.

ENZYMES — Protein substances produced by living cells which can change the rate of chemical reactions. They are organic catalysts.

EROSION — The wearing away of the land surface by detachment and transport of soil and rock materials through the action of moving water, wind or other geological agents.

EVAPOTRANSPIRATION — The loss of water from a soil by evaporation and plant transpiration.

EXCHANGEABLE IONS — Ions held on the soil complex that may be replaced by other ions of like charge. Ions which are held so tightly that they cannot be exchanged are called *nonexchangeable.*

EXCHANGEABLE SODIUM PERCENTAGE — The degree of saturation of the soil exchange complex with sodium. It may be calculated by the formula:

$$ESP = \frac{\text{Exchangeable sodium (me / 100 g soil)}}{\text{Cation exchange capacity (me / 100 g soil)}} \times 100$$

FALLOW — Cropland left idle in order to restore productivity, mainly through accumulation of water, nutrients or both. Summer fallow is a common stage before cereal grain in regions of limited rainfall. The soil is tilled for at least one growing season to control weeds, to aid decomposition of plant residues and to encourage the storage of moisture for the succeeding grain crop. Bush or forest fallow is a rest period under woody vegetation between crops.

FERTILIZER — Any natural or manufactured material added to the soil in order to supply one or more plant nutrients. The term is generally applied to largely inorganic materials other than lime or gypsum sold in the trade.

FERTILIZER FORMULA — The quantity and grade of materials used in making a fertilizer mixture.

FERTILIZER GRADE — An expression that indicates the weight percentage of plant nutrients in a fertilizer. Thus a 10-20-10 grade contains 10 percent nitrogen (N), 20 percent phosphoric acid (P_2O_5) and 10 percent potash (K_2O).

FERTILIZER RATIO — The relative proportions of primary nutrients in a fertilizer grade divided by the highest common divisor for that grade; e.g., grades 10-6-4 and 20-12-8 have the ratio 5-3-2.

FIELD MOISTURE CAPACITY — The moisture content of soil in the field two or three days after a thorough wetting of the soil profile by rain or irrigation water. Field capacity is expressed as moisture percentage, dry-weight basis.

FIFTEEN-ATMOSPHERE PERCENTAGE — The moisture percentage, dry-weight basis, of a soil sample which has been wetted and brought to equilibrium in a pressure-membrane apparatus at a pressure of 221 psi. This characteristic moisture value for soils approximates the lower limit of water available for plant growth. (See also PERMANENT WILTING PERCENTAGE.)

FIXATION — The process by which available plant nutrients are rendered unavailable or "fixed" in the soil. Generally, the process by which potassium, phosphorus and ammonium are rendered unavailable in the soil. Also the process by which free nitrogen is chemically combined either naturally or synthetically. (See also REVERSION and NITROGEN FIXATION.)

FLORICULTURE — Production of foliage or flowering ornamental plants in fields or greenhouses for commercial sales.

FOLIAR FERTILIZATION — Supplying plant nutrients through leaves, with absorption taking place through the stomata of leaves and leaf cuticles.

FORAGE — Unharvested plant material which can be used as feed by domestic animals. Forage may be grazed or cut for hay.

GRADE — The guaranteed analysis of a fertilizer containing one or more of the primary plant nutrient elements. Grades are stated in terms of the guaranteed percentages of nitrogen (N), available phosphate (P_2O_5) and potash (K_2O), in that order. For example, a 10-10-10 grade would contain 10 percent nitrogen, 10 percent available phosphate, and 10 percent potash. (See also ANALYSIS.)

GRAY WATER — See RECYCLED WATER.

GROUND COVER — Plants grown for their low, spreading habit, to protect soils, to prevent the growth of weeds and for aesthetic purposes.

GROWING MEDIA — See POTTING MIX.

GUANO — The decomposed dried excrement of birds and bats, used for fertilizer purposes. The most commonly known guano comes from islands off the coast of Peru and is derived from the excrement of seafowl. It is high in nitrogen and phosphate and at one time was a major fertilizer in this country.

GUARANTEES — The AAPFCO official regulation follows: The statement of guarantees of mixed fertilizer shall be given in whole numbers. All fertilizer components with the exception of potash (K_2O) and phosphoric acid (P_2O_5), if guaranteed, shall be stated in terms of the elements.

GYPSUM ($CaSO_4 \cdot 2H_2O$) — The common name for calcium sulfate, a mineral used in the fertilizer industry as a source of calcium and sulfur. Gypsum also is used widely in reclaiming alkali soils in the western United States. Gypsum cannot be used as a liming material, but it may reduce the alkalinity of sodic soils by replacing sodium with calcium. Another common name is *landplaster*. When pure it contains approximately 18.6 percent sulfur.

HARDPAN — A hardened or cemented soil horizon or layer. The soil material may be sandy or clayey and may be cemented by iron oxide, silica, calcium carbonate or other substances.

HOAGLAND SOLUTION — Nutrient solution containing all essential plant nutrients, and used for hydroponically grown plants. Original nutrient solution was developed by Professor Hoagland at the University of California.

HORIZON, SOIL — A layer of soil, approximately parallel to the soil surface, with distinct characteristics produced by soil-forming processes.

HORTICULTURE — The science of producing and using ornamental plants, fruits and vegetables.

HUMUS — The well-decomposed, more or less stable portion of the organic matter in mineral soils.

HYDROGEN ION CONCENTRATION — See pH.

HYDROPONICS — The production of plants in a liquid solution or gravel medium supplemented with all required nutrients for proper growth.

HYGROSCOPIC — Capable of taking up moisture from the air.

INORGANIC — Substances occurring as minerals in nature or obtainable from them by chemical means. Refers to all matter except the compounds of carbon, but includes carbonates.

INSOLUBLE — Not soluble. As applied to phosphoric acid in fertilizer, that portion of the total phosphoric acid which is soluble neither in water nor in neutral ammonium citrate. As applied to potash and nitrogen, not soluble in water.

ION — An electrically charged particle. As used in soils, an ion refers to an electrically charged element or combination of elements resulting from the breaking up of an electrolyte in solution. Since most soil solutions are very dilute, many of the salts exist as ions. For example, all or part of the potassium chloride (muriate of potash) in most soils exists as potassium ions and chloride ions. The positively charged potassium ion is a cation, and the negatively charged chloride ion is an anion.

KELP — Any of several species of seaweed sometimes harvested for use as a fertilizer. Dried kelp will usually contain 1.6 to 3.3 percent N, 1 to 2 percent P_2O_5 and 15 to 20 percent K_2O.

LEACHING — The removal of materials in solution by the passage of water through soil.

LEACHING REQUIREMENT — The fraction of the water entering the soil that must pass through the root zone in order to prevent soil salinity from exceeding a specified value. Leaching requirement is used primarily under steady-state or long-time average conditions.

LIME — Generally the term *lime*, or *agricultural lime*, is applied to ground limestone (calcium carbonate), hydrated lime (calcium hydroxide) or burned lime (calcium oxide), with or without mixtures of magnesium carbonate, magnesium hydroxide or magnesium oxide, and to materials such as basic slag, used as amendments to reduce the acidity of acid soils. In strict chemical terminology, lime refers to calcium oxide (CaO), but by an extension of meaning it is now used for all limestone-derived materials applied to neutralize acid soils.

LIME REQUIREMENT — The amount of standard ground limestone required to bring a 6.6-inch layer of an acre (about 2 million pounds in mineral soils) of acid soil to some specific lesser degree of acidity, usually to slightly or very slightly acid. In common practice, lime requirements are given in tons per acre of nearly pure limestone, ground finely enough so that all of it passes a 10-mesh screen and at least half of it passes a 100-mesh screen.

LIQUID FERTILIZER — A fluid in which the plant nutrients are in true solution.

LOAM — The textural class name for soil having a moderate amount of sand, silt and clay. Loam soils contain 7 to 27 percent clay, 28 to 50 percent silt, and less than 52 percent sand. (In the old literature, especially English literature, the term *loam* applied to mellow soils rich in organic matter, regardless of the texture. As used in the United States, the term refers only to the relative amounts of sand, silt and clay; loam soils may or may not be mellow.)

LOW-VOLUME IRRIGATION — Irrigation systems including drip, micro-sprinklers, misters or foggers, or any system that is designed to apply water in or near the rooting zone in relatively precise amounts with respect to the plants' needs.

LUXURY CONSUMPTION — The uptake by a plant of an essential nutrient in amounts exceeding what it needs. Thus if potassium is abundant in the soil, alfalfa may take in more than is required.

MACRONUTRIENTS — Nutrients that plants require in relatively large amounts.

MANURE — Generally, the refuse from stables and barnyards, including both animal excreta and straw or other litter. In some other countries the term *manure* is used more broadly and includes both farmyard or animal manure and "chemical manures," for which the term *fertilizer* is nearly always used in the United States.

MARL — An earthy deposit, consisting mainly of calcium carbonate, commonly mixed with clay or other impurities. It is formed chiefly at the margins of freshwater lakes. It is commonly used for liming acid soils.

MICRONUTRIENTS — Nutrients that plants need in only small or trace amounts. Essential micronutrients are boron, chlorine, copper, iron, manganese, molybdenum and zinc.

MILLIEQUIVALENT or MILLIGRAM EQUIVALENT (me) — One-thousandth of an equivalent. In the case of sodium chloride, 1 me would be 0.023 gram of sodium and 0.0355 gram of chloride in 1 liter of water.

MOISTURE RETENTION — The ability of a soil / media to retain moisture. Retentiveness depending upon the type and percentage of materials contained in the soil / media. Generally expressed as a percentage.

MUCK — Highly decomposed organic soil material developed from peat. Generally, muck has a higher mineral or ash content than peat and is decomposed to the point that the original plant parts cannot be identified.

MULCH — A material applied to the ground to prevent excessive drying of the soil surface, to prevent rapid changes in soil temperature, as a soil amendment, for decorative purposes or to prevent weed growth.

MURIATE OF POTASH — Potassium chloride.

NITRIFICATION — The formation of nitrates and nitrites from ammonia (or ammonium compounds), as in soils by microorganisms.

NITROGEN FIXATION — Generally, the conversion of free nitrogen to nitrogen compounds. Specifically in soils, the assimilation of free nitrogen from the soil air by soil organisms and the formation of nitrogen compounds that eventually become available to plants. The nitrogen-fixing organisms associated with legumes are called *symbiotic*; those not definitely associated with the higher plants are *non-symbiotic* or *free-living*.

NONSALINE-SODIC SOIL — A soil which contains sufficient exchangeable sodium to interfere with the growth of most plants but does not contain appreciable quantities of soluble salts. The exchangeable sodium percentage is greater than 15, the conductivity of the saturation extract is less than 4 decisiemens per meter (at 25°C) and the pH of the saturated soil usually ranges between 8.5 and 10.0.

NUTRIENT, PLANT — Any element taken in by a plant which is essential to its growth and which is used by the plant in elaboration of its food and tissue.

NUTRIENT SOLUTION — See HOAGLAND SOLUTION.

ORGANIC — Compounds of carbon other than the inorganic carbonates.

ORGANIC SOIL — A general term applied to a soil or to a soil horizon that consists primarily of organic matter, such as peat soils, muck soils and peaty soil layers.

ORNAMENTAL HORTICULTURE — The branch of horticulture specializing in the areas of floriculture, turfgrass management, nursery stock production and landscaping.

ORTHOPHOSPHATE — A salt of orthophosphoric acid such as ammonium, calcium or potassium phosphate. Each molecule contains a single atom of phosphorus.

ORTHOPHOSPHORIC ACID — H_3PO_4.

PARENT MATERIAL — The unconsolidated mass of rock material (or peat) from which the soil profile develops.

PARTICLE DENSITY — The average density of the soil particles. Particle density is usually expressed in grams per cubic centimeter and is sometimes referred to as *real density* or *grain density.*

PARTS PER MILLION (ppm) — A notation for indicating small amounts of materials. The expression gives the number of units by weight of the substance per million weight units of another substance, such as oven-dry soil. The term may be used to express the number of weight units of a substance per million weight units of a solution. The approximate weight of soil is 2 million pounds per acre – 6 inches. Therefore, ppm $\times$ 2 equals pounds per acre — 6 inches of soil, or ppm $\times$ 4 equals pounds per acre-foot of soil.

PEAT — The AAPFCO has adopted as official the following definition: "Peat is partly decayed vegetable matter of natural occurrence. It is composed chiefly of organic matter that contains some nitrogen of low activity."

PERCOLATION — The downward movement of water through soil.

PERENNIAL PLANT — A plant that lives for more than two years.

PERMANENT WILTING PERCENTAGE — The moisture percentage of soil at which plants wilt and fail to recover turgidity (15 atmospheres). It is usually determined with dwarf sunflowers. The expression has significance only for non-saline soils.

PERMEABILITY, SOIL — The quality of a soil horizon that enables water or air to move through it. It can be measured quantitatively in terms of rate of flow of water through a unit cross section in unit time under specified temperature and hydraulic conditions. Values for saturated soils usually are called *hydraulic conductivity.* The permeability of a soil is controlled by the least permeable horizon even though the others are permeable.

pH — A numerical designation of acidity and alkalinity as in soils and other biological systems. Technically, pH is the common logarithm of the reciprocal of the hydrogen ion concentration of a solution. A pH of 7.0 indicates precise neutrality; higher values indicate increasing alkalinity, and lower values indicate increasing acidity.

PHOSPHATE — A salt of phosphoric acid made by combining phosphoric acid with ions such as ammonium, calcium, potassium or sodium.

PHOSPHATE ROCK — Phosphate-bearing ore composed largely of tricalcium phosphate. Phosphate rock can be treated with strong acids or heat to make available forms of phosphate. Finely ground rock phosphate is sometimes used in long-time fertility programs.

PHOSPHORIC ACID — A term that refers to the phosphorus content of a fertilizer, expressed as phosphoric acid (P_2O_5). The AAPFCO has adopted as official the following definition: "The term phosphoric acid designates P_2O_5." Phosphoric acid also refers to the acid H_3PO_4.

PHOTOSYNTHESIS — The process by which green plants combine water and carbon dioxide to form carbohydrates under the action of light. Chlorophyll is required for the conversion of light energy into chemical energy.

POLYPHOSPHATE — A salt of polyphosphoric acid such as ammonium, calcium or potassium polyphosphate. *Poly* means "many" and refers to multiple linkages of phosphorus in each molecule.

POLYPHOSPHORIC ACID — Condensed phosphoric acid ranging in P_2O_5 content from 68 to 83 percent.

POROSITY — The fraction of soil volume not occupied by soil particles.

POTASH — The AAPFCO has adopted as official the following definition: "The term potash designates potassium oxide (K_2O)."

POTTING MIX — A mixture of various organic and inorganic constituents, including soil, which is used for growing plants in containers or beds.

PRIMARY PLANT NUTRIENTS (plant foods) — Nitrogen (N), phosphate (P_2O_5) and potash (K_2O).

PRODUCTIVITY — In simplest terms, the ability of the soil to produce. It differs from fertility to the extent that a soil may be fertile and yet unable to produce because of other limiting factors.

PROFILE, SOIL — A vertical section of the soil extending through all its horizons and into the parent material.

PROTEIN — Any of a group of high-molecular-weight nitrogen-containing compounds that yield amino acids on hydrolysis. Protein is a vital part of living matter and is one of the essential food substances of animals.

PROTOPLASM — The basic, jellylike substance in plant and animal cells; it is basic to all life processes.

PUDDLED SOIL — Dense, massive soil artificially compacted when wet and having no regular structure. The condition commonly results from tillage of or heavy traffic on a clayey soil when it is wet.

QUICK TESTS — Simple and rapid chemical tests of soils designed to give an approximation of the nutrients available to plants.

RATIO — See FERTILIZER RATIO.

RECLAMATION — The process of restoring lands to productivity by removing excess soluble salts or excess exchangeable sodium from soils.

RECYCLED WATER — Water that has been used for domestic, industrial, or agricultural purposes, but not exposed to extensive treatment prior to reuse.

REVERSION — The interaction of a plant nutrient with the soil which causes the nutrient to become less available. In fertilizer manufacturing, the excessive use of ammonia in ammoniation of phosphates results in phosphate reversion. (See also FIXATION.)

ROOTBOUND — Having a closely packed mass of roots, as on a plant that has grown too large for its container.

SALINE-SODIC SOIL / MEDIA — A soil / media containing sufficient exchangeable sodium to interfere with the growth of most crop plants and containing appreciable quantities of soluble salts. The exchangeable sodium percentage is greater than 15, and the electrical conductivity of the saturation extract is greater than 4 decisiemens per meter (at 25°C). The pH reading of the saturated soil is usually less than 8.5.

SALINE SOIL / MEDIA — A soil / media containing enough soluble salts to impair its productivity for plants but not containing an excess of exchangeable sodium.

SALT INDEX — An index used to compare solubilities of chemical compounds. Most nitrogen and potash compounds have high indexes, and phosphate compounds have low indexes. When applied too close to seed or on foliage, the compounds with high indexes cause plants to wilt or die.

SALTING OUT — Precipitation of the dissolved salts contained in a solution when the temperature drops to a certain point (the salting out point).

SALTS — The products, other than water, of the reaction of an acid with a base. Salts commonly found in soils break up into cations (sodium, calcium, etc.) and anions (chloride, sulfate, etc.) when dissolved in water.

SAND — Individual rock or mineral fragments in soils having diameters ranging from 0.05 millimeter to 2.0 millimeters. Usually sand grains consist chiefly of quartz, but they may be of any mineral composition. The textural class name of any soil that contains 85 percent or more sand and not more than 10 percent clay.

SATURATED SOIL PASTE — A particular mixture of soil and water commonly used for measurements and for obtaining soil extracts. At saturation the soil paste glistens as it reflects light, flows slightly when the container is tipped and slides freely and cleanly from a spatula for all soils except those with high clay content.

SATURATION EXTRACT — The solution extracted from a soil at its saturation percentage.

SATURATION PERCENTAGE — The moisture percentage of a saturated soil paste, expressed on a dry-weight basis.

SECONDARY PLANT NUTRIENTS — Calcium, magnesium and sulfur.

SEPARATE, SOIL — One of the individual-size groups of mineral soil particles — sand, silt or clay.

SERIES, SOIL — A group of soils that have soil horizons similar in their differentiating characteristics and arrangement in the soil profile, except for the texture of the surface soil, and are formed from a particular type of parent material. Soil series is an important category in detailed soil classification. Individual series are given proper names from place names near the first recorded occurrence. Thus, Yolo, Panoche, Hanford and San Joaquin are names of soil series that appear on soil maps, and each denotes a unique combination of many soil characteristics.

SEWAGE SLUDGE — An organic product resulting from the treatment of sewage. The composition varies widely depending on the method of treatment.

SILT — (1) Individual mineral particles of soil that range in diameter between the upper size of clay, 0.002 mm, and the lower size of very fine sand, 0.05 mm. (2) Soil of the textural class silt containing 80 percent or more silt and less than 12 percent clay. (3) Sediments deposited from water in which the individual grains are approximately the size of silt, although the term is sometimes applied loosely to sediments containing considerable sand and clay.

SILVICULTURE — A branch of forestry dealing with the development and care of forests.

SLURRY FERTILIZER — A fluid mixture containing dissolved and undissolved plant nutrient materials which requires continuous mechanical agitation to assure homogeneity.

SODIC SOIL / MEDIA — A soil / media that contains sufficient exchangeable sodium to interfere with the growth of most plants, either with or without appreciable quantities of soluble salts. (See also NONSALINE-SODIC SOIL and SALINE-SODIC SOIL / MEDIA.)

SODIUM ADSORPTION RATIO — A ratio for soil extracts and irrigation waters used to express the relative activity of sodium ions in exchange reactions with soil.

$$ SAR = \frac{Na^+}{\sqrt{(Ca^{++} + Mg^{++})/2}} $$

The ionic concentrations are expressed in milliequivalents per liter.

SODIUM PERCENTAGE — The percent sodium of total cations. Calculations are based on milliequivalents rather than weight.

SOIL MOISTURE STRESS — The sum of the soil moisture tension and the osmotic pressure of the soil solution. It is the force plants must overcome to withdraw moisture from the soil.

SOIL MOISTURE TENSION — The force by which moisture is held in the soil. It is a negative pressure and may be expressed in any convenient pressure unit. Tension does not include osmotic pressure values.

STRUCTURE, SOIL — The physical arrangement of the soil particles.

SUBSOIL — Roughly, that part of the soil below plow depth.

SUB-SURFACE IRRIGATION — The distribution of irrigation water through conduits below the surface of the soil. Definition may also apply to water available to plants from a water table.

SUPERPHOSPHATE — The AAPFCO has adopted as official the following definition: "Superphosphate is a product obtained by mixing rock phosphate with either sulfuric acid or phosphoric acid or with both acids. (The grade that shows the available phosphoric acid shall be used as a prefix to the name. Example: 20 percent superphosphate.)"

SUPERPHOSPHORIC ACID — See POLYPHOSPHORIC ACID.

SUSPENSION FERTILIZER — A fluid containing dissolved and undissolved plant nutrients. The suspension of the undissolved plant nutrients may be inherent to the materials or produced with the aid of a suspending agent of non-fertilizer properties. Mechanical agitation may be necessary in some cases to facilitate uniform suspension of undissolved plant nutrients.

SYMBIOSIS — The living together of two different organisms with a resulting mutual benefit. A common example is the association of rhizobia with legumes; the resulting nitrogen fixation is sometimes called *symbiotic nitrogen fixation*. Adjective: symbiotic.

TANKAGE — Dried animal residue. Process tankage is made from leather scrap, wool and other inert nitrogenous materials by steaming under pressure with or without addition of acid. This treatment increases the availability of the nitrogen to plants.

TENSIOMETER — A device used to measure the tension with which water is held in the soil.

TEXTURE, SOIL / MEDIA — The relative proportions of the various size groups of individual soil grains in a mass of soil / media. Specifically, it refers to the proportions of sand, silt and clay.

TILTH — The physical condition of a soil with respect to its fitness for the growth of plants.

TRACE ELEMENTS — See MICRONUTRIENTS.

TRANSPIRATION — Loss of water vapor from the leaves and stems of living plants to the atmosphere.

TRIPLE SUPERPHOSPHATE — A product that contains 40 to 50 percent available phosphoric acid. Triple superphosphate differs from ordinary superphosphate in that it contains very little calcium sulfate. In the fertilizer trade, the product is also called *treble superphosphate, concentrated superphosphate, double super-phosphate* and *multiple superphosphate.*

TURF — Any grassy area maintained by frequent mowing, fertilization and watering used for lawns, roadsides or playing fields.

TURFGRASS — A species or cultivar of grass, usually of spreading habit, which is maintained as a mowed turf.

UNCOATED ORGANIC COMPOUNDS — Generally, organic nitrogen compounds, usually based on urea, that release nitrogen slowly over given periods of time.

UNIT — The AOAC has adopted as official the following definition: "A unit of plant food is twenty (20) pounds, or one percent (1 percent) of a ton."

VOLATILIZATION — The evaporation or changing of a substance from liquid to vapor.

WATER-HOLDING CAPACITY — The weight of water held by a given quantity of absolutely dry soil / media when saturated.

WATER TABLE — The upper surface of ground water.

WATER TABLE, PERCHED — The upper surface of a body of free ground water in a zone of saturation separated from underlying ground water by unsaturated material.

WEATHERING — The physical and chemical disintegration and decomposition of parent material as in soil formation.

WILTING PERCENTAGE — See PERMANENT WILTING PERCENTAGE.

WINDBREAKS — A group of plants placed in locations where they might screen out winds and snow drifting around a house or landscape.

WINTER HARDINESS — The plant characteristic of being able to successfully withstand the rigors of winter.

APPENDIX B

Model Law Relating to Fertilizing Materials

The Association of American Plant Food Control Officials was formed in 1946. The objectives of the Association are to promote uniform and effective legislation, definitions, rulings and enforcement of laws relating to the control of sale and distribution of mixed fertilizers on the continent of North America. These officials function to protect the fertilizer industry and the farmer alike against any who seek to practice unfair dealings.

The membership of the association consists of:

1. Officers charged by law with the active execution of the laws regulating the sale of commercial fertilizer and fertilizer materials.
2. Deputies who may be designated by the officials named in Section 1.
3. Research workers employed by state, dominion or federal agencies who are engaged in the investigation of fertilizers.

This organization studies through committees or investigation any problems which may arise concerning the various points listed as objectives. Regular meetings are held at least once a year. Through their association, the fertilizer control officers have been able to work for improved laws and definitions and to discuss mutual problems of enforcement.

During recent years a proposed "Model State Fertilizer Bill" was drafted in an effort to bring about uniformity in fertilizer regulation. The provisions of the fertilizer laws in most states and provinces follow this "Model Bill." However, regulations covering registration, labeling, reporting of tonnage, violations and tonnage fees vary considerably in different areas. It is recommended that those desiring copies of laws which apply to the particular area in which they operate write to the governing fertilizer control official. Addresses are given below.

For information about states not listed in this chapter, write to the Association of American Plant Food Control Officials, Department of Biochemistry, Purdue University, West Lafayette, IN 47907. The association publishes annually an official publication which contains the constitution and bylaws, fertilizer terms, annual report and a complete list of state fertilizer control officials.

Alaska
  Division of Agriculture
  Department of Natural Resources
  P.O. Box 949
  Palmer, AK 99645-0949

Arizona
  Office of State Chemist
  State Agricultural Laboratory
  P.O. Box 1586
  Mesa, AZ 85201

233

California
Feed, Fertilizer & Livestock Drug
Unit
1220 N Street — Room A-372
Sacramento, CA 95814

Canada
Feed and Fertilizer Division
Plant Health & Plant Products
Directorate
K. W. Neatby Building
Central Experimental Farm
Ottawa, Ontario K1A 0C6
Canada

Colorado
Department of Agriculture
2331 West 31st Avenue
Denver, CO 80211

Hawaii
Chief, Commodities Branch
Department of Agriculture
P.O. Box 5425
Honolulu, HI 96814

Idaho
Bureau of Feed & Fertilizer
Control
Department of Agriculture
P.O. Box 790
Boise, ID 83701

Montana
Plant Industry Division
State Department of Agriculture
Helena, MT 59601

Nevada
Chief Chemist
Department of Agriculture
P.O. Box 11100
Reno, NV 89510

New Mexico
Chief, Division, Feed, Seed &
Fertilizer
P.O. Box 3150
Las Cruces, NM 88003

Oregon
Administrator, Plant Division
Department of Agriculture
635 Capitol Street, N.E.
Salem, OR 97310

Utah
Division of Plant Industry
Department of Agriculture
350 North Redwood Road
Salt Lake City, UT 84116

Washington
Agricultural Chemical & Plant
Services Division
Department of Agriculture
406 General Administration
Building
Olympia, WA 98504

Wyoming
Director, Division of Plant Industry
2219 Carey Avenue
Cheyenne, WY 82002

OFFICIALLY ADOPTED DOCUMENTS

Note — Although these documents have not been passed into law in all states, the subject matter covered herein does represent the official policy of this Association.

UNIFORM STATE FERTILIZER BILL
(Official 1982)

AN ACT to regulate the sale and distribution of fertilizers in the State of _____. BE IT ENACTED by the legislature of the State of _____.

Section 1. Title

    This Act shall be known as the "_____ Fertilizer Law of 19_____".

Section 2. Enforcing Official

    This Act shall be administered by the _____ of the State of _____, hereinafter referred to as the "_____".

Section 3. Definitions of Words and Terms

    When used in this Act:

(a) The term "fertilizer" means any substance containing one or more recognized plant nutrient(s) which is used for its plant nutrient content and which is designed for use or claimed to have value in promoting plant growth, except unmanipulated animal and vegetable manures, marl, lime, limestone, wood ashes and other products exempted by regulation by the _____.

    (1) A "fertilizer material" is a fertilizer which either:
        A. Contains important quantities of no more than one of the primary plant nutrients: nitrogen (N), phosphorus (P) and potassium (K), or
        B. Has 85 percent or more of its plant nutrient content present in the form of a single chemical compound, or
        C. Is derived from a plant or animal residue or by-product or natural material deposit which has been processed in such a way that its content of plant nutrients has not been materially changed except by purification and concentration.
    (2) A "mixed fertilizer" is a fertilizer containing any combination or mixture of fertilizer materials.
    (3) A "specialty fertilizer" is a fertilizer distributed for non-farm use.
    (4) A "bulk fertilizer" is a fertilizer distributed in a non-packaged form.

(b) The term "brand" means a term, design, or trademark used in connection with one or several grades of fertilizer.

(c) Guaranteed Analysis:
    (1) Until the _____ prescribes the alternative form of "Guaranteed Analysis" in accordance with the provisions of subparagraph (2) hereof, the term "Guaranteed Analysis" shall mean the minimum

percentage of plant nutrients claimed in the following order and form:

A. Total Nitrogen (N)   _____ per cent
    Available Phosphoric Acid (P_2O_5)   _____ per cent
    Soluble Potash (K_2O)   _____ per cent
B. For unacidulated mineral phosphatic material and basic slag, bone, tankage and other organic phosphatic materials, the total phosphoric acid and / or degree of fineness may also be guaranteed.
C. Guarantees for plant nutrients other than nitrogen, phosphorus and potassium may be permitted or required by regulation by the _____. The guarantees for such other nutrients shall be expressed in the form of the element. The source (oxides, salts, chelates, etc.) of such other nutrients may be required to be stated on the application for registration and may be included on the label. Other beneficial substances or compounds, determinable by laboratory methods, also may be guaranteed by permission of the _____ and with the advice of the Director of the Agricultural Experiment Station. When any plant nutrients or other substances or compounds are guaranteed, they shall be subject to inspection and analysis in accord with the methods and regulations prescribed by the _____.

(2) When the _____ finds, after public hearing following due notice, that the requirement for expressing the guaranteed analysis of phosphorus and potassium in elemental form would not impose an economic hardship on distributors and users of fertilizer by reason of conflicting labeling requirements among the states, he may require by regulation thereafter that the "Guaranteed Analysis" shall be in the following form:

Total Nitrogen (N)   _____ per cent
Available Phosphorus (P)   _____ per cent
Soluble Potassium (K)   _____ per cent

Provided, however, That the effective date of said regulation shall be not less than six months following the issuance thereof, and Provided, further, That for a period of two years following the effective date of said regulation the equivalent of phosphorus and potassium may also be shown in the form of phosphoric acid and potash; Provided, however, That after the effective date of a regulation issued under the provisions of this section, requiring that phosphorus and potassium be shown in the elemental form, the guaranteed analysis for nitrogen, phosphorus, and potassium shall constitute the grade.

(d) The term "grade" means the percentage of total nitrogen, available phosphorus or phosphoric acid, and soluble potassium or potash stated in whole numbers in the same terms, order, and percentages as in the guaranteed analysis. Provided, however, That specialty fertilizers may be guaranteed in fractional units of less than one per cent of total nitrogen, available phosphorus or phosphoric acid, and soluble potassium or potash: Provided, further, That fertilizer materials, bone meal, manures, and similar materials may be guaranteed in fractional units.

(e) The term "official sample" means any sample of fertilizer taken by the _____ or his agent and designated as "official" by the _____.

(f) The term "ton" means a net weight of two thousand pounds avoirdupois.

(g) The term "primary nutrient" includes nitrogen, available phosphoric acid or phosphorus, and soluble potash or potassium.

(h) The term "per cent" or "percentage" means the percentage by weight.

(i) The term "person" includes individual, partnership, association, firm and corporation.

(j) The term "distribute" means to import, consign, manufacture, produce, compound, mix, or blend fertilizer, or to offer for sale, sell, barter or otherwise supply fertilizer in this state.

(k) The term "distributor" means any person who distributes.

(l) The term "registrant" means the person who registers fertilizer under the provisions of this Act.

(m) The term "licensee" means the person who receives a license to distribute a fertilizer under the provisions of this Act.

(n) The term "label" means the display of all written, printed, or graphic matter, upon the immediate container, or a statement accompanying a fertilizer.

(o) The term "labeling" means all written, printed, or graphic matter, upon or accompanying any fertilizer, or advertisements, brochures, posters, television and radio announcements used in promoting the sale of such fertilizer.

(p) The term "investigational allowance" means an allowance for variations inherent in the taking, preparation and analysis of an official sample of fertilizer.

(q) The term "deficiency" means the amount of nutrient found by analysis less than that guaranteed which may result from a lack of nutrient ingredients or from lack of uniformity. (Official 1985.)

Section 4. Option A — Registration

(a) Each brand and grade of fertilizer shall be registered in the name of that person whose name appears upon the label before being distributed in this state. The application for registration shall be submitted to the _____ on a form furnished by the _____ and shall be accompanied by a fee of $_____ per each grade of each brand except those fertilizers sold in packages of 10 pounds or less shall be registered at a fee of $_____ per each grade of each brand. Upon approval by the _____ a copy of the registration shall be furnished to the applicant. All registrations expire on _____ each year. The application shall include the following information:
   (1) The brand and grade;
   (2) The guaranteed analysis;
   (3) The name and address of the registrant;
   (4) Net weight.

(b) A distributor shall not be required to register any fertilizer which is already registered under this Act by another person, providing the label does not differ in any respect.

(c) A distributor shall not be required to register each grade of fertilizer formulated according to specifications which are furnished by a consumer prior to mixing, but shall be required to register his firm in a manner and at a fee as prescribed in regulations by the _____ and to label such fertilizer as provided in Section 5(b).

Section 4. Option B — Registration and Licensing

(a) No person whose name appears upon the label of a fertilizer shall distribute that fertilizer, except specialty fertilizers, to a non-licensee until a license to distribute has been obtained by that person from the _____ upon payment of a $_____ fee. All licenses expire on the _____ day of _____ each year.

(b) An application for license shall include:
   (1) The name and address of licensee.
   (2) The name and address of each distribution point in the state. The name and address shown on the license shall be shown on all labels, pertinent invoices, and storage facilities for fertilizer distributed by the licensee in this state.

(c) The licensee shall inform the _____ in writing of additional distribution points established during the period of the license.

(d) No person shall distribute in this state a specialty fertilizer until it is registered with the _____ by the distributor whose name appears on the label. An application for each brand and product name of each grade of specialty fertilizer shall be made on a form furnished by the _____ and shall be accompanied by a fee of $_____ per each grade of each brand, except those fertilizers sold in packages of 10 pounds or less shall be registered at a fee of $_____ per each grade of each brand. Labels for each brand and product name of each grade shall accompany the application. Upon the approval of an application by the _____, a copy of the registration shall be furnished the applicant. All registrations expire on the _____ day of _____ each year.

(e) An application for registration shall include the following:
   (1) The brand and grade;
   (2) The guaranteed analysis;
   (3) Name and address of the registrant;
   (4) Net weight.

Section 4. Option C — Licensing

(a) No person whose name appears upon the label of a fertilizer shall distribute that fertilizer to a non-licensee until a license to distribute has been obtained by that person from the _____ upon payment of a $_____ fee. All licenses expire on the _____ day of each year.

(b) An application for license shall include:
   (1) The name and address of licensee.
   (2) The name and address of each distribution point in the state. The name and address shown on the license shall be shown on all labels, pertinent invoices, and storage facilities for fertilizers distributed by the licensee in this state.

(c) The licensee shall inform the _____ in writing of additional distribution points established during the period of the license.

Section 5. Labels
- (a) Any fertilizer distributed in this state in containers shall have placed on or affixed to the container a label setting forth in clearly legible and conspicuous form the following information:
  - (1) Net weight;
  - (2) Brand and grade: Provided, That the grade shall not be required when no primary nutrients are claimed;
  - (3) Guaranteed analysis;
  - (4) Name and address of the registrant / licensee.

  In case of bulk shipments, this information in written or printed form shall accompany delivery and be supplied to the purchaser at time of delivery.
- (b) A fertilizer formulated according to specifications which are furnished by / for a consumer prior to mixing shall be labeled to show the net weight, the guaranteed analysis, and the name and address of the distributor or registrant / licensee.

Section 6. Inspection Fees
- (a) There shall be paid to the _____ for all fertilizers distributed in this state to non-registrants / non-licensees an inspection fee at the rate of _____ cents per ton; Provided, That sales or exchanges between importers, manufacturers, distributors or registrants / licensees are hereby exempted.
- (b) Every registrant / licensee who distributes fertilizer in the state shall file with the _____ a (monthly, quarterly, or semi-annual) statement for the reporting period setting forth the number of net tons of each fertilizer so distributed in this state during such period. The report shall be due on or before thirty days following the close of the filing period and upon such statement shall pay the inspection fee at the rate stated in paragraph (a) of this section. If the tonnage report is not filed and the payment of inspection fees is not made within 30 days after the end of the specified filing period, a collection fee, amounting to 10 per cent (minimum $10) of the amount due, shall be assessed against the registrant / licensee and added to the amount due.
- (c) When more than one person is involved in the distribution of a fertilizer, the last person who has the fertilizer registered (is licensed) and who distributed to a non-registrant / licensee dealer, or consumer is responsible for reporting the tonnage and paying the inspection fee, unless the report and payment is made by a prior distributor of the fertilizer.
- (d) On individual packages of fertilizer containing 10 pounds or less there shall be paid, in lieu of the inspection fee of _____ cents per ton and in lieu of $_____ per brand and grade, an annual registration and inspection fee of $_____ for each grade of each brand sold or distributed. Where a person distributes fertilizer in packages of 10 pounds or less and in packages over 10 pounds, the annual fee shall apply only to that portion distributed in packages of 10 pounds or less.
- (e) Fees so collected shall be used for the payment of the costs of inspection, sampling and analysis, and other expenses necessary for the administration of this Act.

Section 7. Tonnage Reports

(a) The person distributing or selling fertilizer to a non-registrant / non-licensee shall furnish the _____ a report showing the county of the consignee, the amounts (tons) of each grade of fertilizer, and the form in which the fertilizer was distributed (bags, bulk, liquid, etc.). This information shall be reported by one of the following methods:

(1) Submitting a summary report approved by the _____ on or before the _____ day of each month covering shipments made during the preceding month; or

(2) Submitting a copy of the invoice within _____ business days after shipment.

(b) No information furnished the _____ under this section shall be disclosed in such a way as to divulge the operation of any person.

Section 8. Inspection, Sampling, Analysis

(a) It shall be the duty of the _____, who may act through his authorized agent, to sample, inspect, make analyses of, and test fertilizers distributed within this state at any time and place and to such an extent he may deem necessary to determine whether such fertilizers are in compliance with the provisions of this Act. The _____, individually or through his agent, is authorized to enter upon any public or private premises or carriers during regular business hours in order to have access to fertilizer subject to provisions of this Act and the regulations pertaining thereto, and to the records relating to their distribution.

(b) The methods of sampling and analysis shall be those adopted by the Association of Official Analytical Chemists (AOAC). In cases not covered by such methods, or in cases where methods are available in which improved applicability has been demonstrated, the _____ may adopt such appropriate methods from other sources.

(c) The _____, in determining for administrative purposes whether any fertilizer is deficient in plant food, shall be guided solely by the official sample as defined in paragraph (e) of Section 3, and obtained and analyzed as provided for in paragraph (b) of this section.

(d) The results of official analysis of fertilizers and portions of official samples shall be distributed by the _____ as provided by regulation. Official samples establishing a penalty for nutrient deficiency shall be retained for a minimum of 90 days from issuance of a deficiency report.

Section 9. Plant Food Deficiency

(a) Penalty for nitrogen, available phosphoric acid or phosphorus, and soluble potash or potassium — If the analysis shall show that a fertilizer is deficient (1) in one or more of its guaranteed primary plant nutrients beyond the investigational allowances and compensations as established by regulation, or (2) if the overall index value of the fertilizer is below the level established by regulation, a penalty payment of _____ times the value of such deficiency or deficiencies shall be assessed. When a fertilizer is subject to a penalty payment under both (1) and (2), the larger penalty payment shall apply.

(b) Penalty payment for other deficiencies — Deficiencies beyond the investigational allowances as established by regulation in any other constituent(s) covered under Section 3 paragraph (c)(1) B and C of this Act, which the registrant / licensee is required to or may guarantee, shall be evaluated and penalty payments prescribed by the _____.

(c) All penalty payments assessed under this section shall be paid by the registrant / licensee to the consumer of the lot of fertilizer represented by the sample analyzed within three months after the date of notice from the _____ to the registrant / licensee, receipts taken therefor and promptly forwarded to the _____. If said consumer cannot be found, the amount of the penalty payments shall be paid to the _____ who shall deposit the same in the appropriate state fund allocated to fertilizer control service. If upon satisfactory evidence a person is shown to have altered the content of a fertilizer shipped to him by a registrant / licensee, or to have mixed or commingled fertilizer from two or more suppliers such that the result of either alteration changes the analysis of the fertilizer as originally guaranteed, then that person shall become responsible for obtaining a registration / license and shall be held liable for all penalty payments and be subject to other provisions of this Act, including seizure, condemnation and stop sale.

(d) A deficiency in an official sample of mixed fertilizer resulting from non-uniformity is not distinguishable from a deficiency due to actual plant nutrient shortage and is properly subject to official action. (Official 1985.)

(e) Nothing contained in this section shall prevent any person from appealing to a court of competent jurisdiction praying for judgment as to the justification of such penalty payments.

Section 10. Commercial Value

For the purpose of determining the commercial value to be applied under the provisions of Section 9 the _____ shall determine and publish annually the values per unit of nitrogen, available phosphoric acid, and soluble potash in fertilizers in this state. If guarantees are as provided in Section 3(c)(2) the value shall be per unit of nitrogen, phosphorus and potassium. The value so determined and published shall be used in determining and assessing penalty payments.

Section 11. Misbranding

No person shall distribute misbranded fertilizer. A fertilizer shall be deemed to be misbranded:

(a) If its labeling is false or misleading in any particular.

(b) If it is distributed under the name of another fertilizer product.

(c) If it is not labeled as required in Section 5 of this Act and in accordance with regulations prescribed under this Act.

(d) If it purports to be or is represented as a fertilizer, or is represented as containing a plant nutrient or fertilizer unless such plant nutrient or fertilizer conforms to the definition of identity, if any, prescribed by regulation of the _____; in adopting such regulations the _____ shall give due regard to commonly accepted definitions and official fertilizer terms such as those issued by the Association of American Plant Food Control Officials.

Section 12. Adulteration

No person shall distribute an adulterated fertilizer product. A fertilizer shall be deemed to be adulterated:

(a) If it contains any deleterious or harmful ingredient in sufficient amount to render it injurious to beneficial plant life when applied in accordance with directions for use on the label, or if adequate warning statements or directions for use which may be necessary to protect plant life are not shown upon the label.

(b) If its composition falls below or differs from that which it is purported to possess by its labeling.

(c) If it contains unwanted crop seed or weed seed.

Section 13. Publications

The _____ shall publish at least annually and in such forms as he may deem proper: (a) Information concerning the distribution of fertilizers, (b) Results of analyses based on official samples of fertilizer distributed within the state as compared with analyses guaranteed under Section 4 and Section 5.

Section 14. Rules and Regulations

The _____ is authorized to prescribe and, after a public hearing following due public notice, to enforce such rules and regulations relating to investigational allowances, definitions, records, and the distribution of fertilizers as may be necessary to carry into effect the full intent and meaning of this Act.

Section 15. Short Weight

If any fertilizer in the possession of the consumer is found by the _____ to be short in weight, the registrant / licensee of said fertilizer shall within thirty days after official notice from the _____ submit to the consumer a penalty payment of _____ times the value of the actual shortage.

Section 16. Cancellation of Registration / License

The _____ is authorized and empowered to cancel the registration (license of any person) of any brand of fertilizer or to refuse to register any brand of fertilizer (issue a license) as herein provided, upon satisfactory evidence that the registrant / licensee has used fraudulent or deceptive practices in the evasion or attempted evasion of the provisions of this Act or any regulations promulgated thereunder: Provided, That no license or registration shall be revoked or refused until the person (registrant / licensee) shall have been given the opportunity to appear for a hearing by the _____.

Section 17. "Stop Sale" Orders

The _____ may issue and enforce a written or printed "stop sale, use, or removal" order to the owner or custodian of any lot of fertilizer and to hold at a designated place when the _____ finds said fertilizer is being offered or exposed for sale in violation of any of the provisions of this Act until the law has been complied with and said fertilizer is released in writing by the _____, or said violation has been otherwise legally disposed of by written authority. The _____ shall release the fertilizer so withdrawn when the requirements of the provisions of this Act have been complied with and all costs and expenses incurred in connection with the withdrawal have been paid.

Section 18. Seizure, Condemnation and Sale

Any lot of fertilizer not in compliance with the provisions of this Act shall be subject to seizure on complaint of the _____ to a court of competent jurisdiction in the area in which said fertilizer is located. In the event the court finds the said fertilizer to be in violation of this Act and orders the condemnation of said fertilizer it shall be disposed of in any manner consistent with the quality of the fertilizer and the laws of the state: Provided, That in no instance shall the disposition of said fertilizer be ordered by the court without first giving the claimant an opportunity to apply to the court for release of said fertilizer or for permission to process or relabel said fertilizer to bring it into compliance with this Act.

Section 19. Violations

(a) If it shall appear from the examination of any fertilizer that any of the provisions of this Act or the rules or regulations issued thereunder have been violated, the _____ shall cause notice of the violations to be given to the registrant / licensee, distributor, or processor from whom said sample was taken; any person so notified shall be given opportunity to be heard under such regulations as may be prescribed by the _____. If it appears after such hearing, either in the presence or absence of the person so notified, that any of the provisions of this Act or rules and regulations issued thereunder have been violated, the _____ may certify the facts to the proper prosecuting attorney.

(b) Any person convicted of violating any provision of this Act or the rules and regulations issued thereunder shall be punished in the discretion of the court.

(c) Nothing in this Act shall be construed as requiring the _____ or his representative to report for prosecution or for the institution of seizure proceedings as a result of minor violations of the Act when he believes that the public interests will be best served by a suitable notice of warning in writing.

(d) It shall be the duty of each _____ attorney to whom any violation is reported to cause appropriate proceedings to be instituted and prosecuted in a court of competent jurisdiction without delay.

(e) The _____ is hereby authorized to apply for and the court to grant a temporary or permanent injunction restraining any person from violating or continuing to violate any of the provisions of this Act or any rule or regulation promulgated under this Act notwithstanding the existence of other remedies at law. Said injunction to be issued without bond.

Section 20. Exchanges Between Manufacturers

Nothing in this Act shall be construed to restrict or avoid sales or exchanges of fertilizers to each other by importers, manufacturers, or manipulators who mix fertilizer materials for sale, or as preventing the free and unrestricted shipments of fertilizer to manufacturers or manipulators who have registered their brands (are licensed) as required by provisions of this Act.

Section 21. Constitutionality

If any clause, sentence, paragraph or part of this Act shall for any reason be judged invalid by any court of competent jurisdiction, such judgment shall not affect, impair, or invalidate the remainder thereof but shall be confined in its operation to the clause, sentence, paragraph, or part thereof directly involved in the controversy in which such judgment shall have been rendered.

Section 22. Repeal

All laws and parts of laws in conflict with or inconsistent with the provisions of this Act are hereby repealed.

Section 23. Effective Date

This Act shall take effect and be in force from and after the first day of _____.

RULES AND REGULATIONS

Under the Uniform Fertilizer Bill by the _____ of the State of _____ Pursuant to due publication and notice of opportunity for a public hearing, the _____ has adopted the following regulations.

1. Plant Nutrients in Addition to Nitrogen, Phosphorus and Potassium.

Other plant nutrients when mentioned in any form or manner shall be registered and shall be guaranteed. Guarantees shall be made on the elemental basis. Sources of the elements guaranteed and proof of availability shall be provided the _____ upon request. Except guarantees for those water soluble nutrients labeled for hydroponic or continuous liquid feed programs, the minimum percentages which will be accepted for registration are as follows:

Element	%
Calcium (Ca)	1.0000
Magnesium (Mg)	0.5000
Sulfur (S)	1.0000
Boron (B)	0.0200
Chlorine (Cl)	0.1000
Cobalt (Co)	0.0005
Copper (Cu)	0.0500
Iron (Fe)	0.1000
Manganese (Mn)	0.0500
Molybdenum (Mo)	0.0005
Sodium (Na)	0.1000
Zinc (Zn)	0.0500

Guarantees or claims for the above listed plant nutrients are the only ones which will be accepted. Proposed labels and directions for the use of the fertilizer shall be furnished with the application for registration upon request. Any of the above listed elements which are guaranteed shall appear in the order listed immediately following guarantees for the primary nutrients of nitrogen, phosphorus and potassium. (Official 1987.)

[Secretary's Note — Paragraphs 3 and 4 (Off. Publication No. 38) were deleted — Official 1985]

A warning or caution statement may be required for any product which contains *(name of micro-nutrient)* in water soluble form when there is evidence that *(name of micro-nutrient)* in excess of _____ % may be harmful to certain crops or where there are unusual environmental conditions. (Official 1984.)

Examples of Warning or Caution Statements

Boron:
1. Directions: Apply this fertilizer at a maximum rate of 350 pounds per acre for Alfalfa or Red Clover seed production. CAUTION: Do not use on other crops. The Boron may cause injury to them.
2. CAUTION: Apply this fertilizer at a maximum rate of 700 pounds per acre for Alfalfa or Red Clover seed production. Do not use on other crops; the boron may cause serious injury to them.
3. WARNING: This fertilizer carries added borax and is intended for use only on Alfalfa. Its use on any other crops or under conditions other than those recommended may result in serious injury to the crops.

Molybdenum:
1. CAUTION: This fertilizer is to be used only on soil which responds to molybdenum. Crops high in molybdenum are toxic to grazing animals (ruminants).

2. Fertilizer Labels.

The following information, in the format presented, is the minimum required for all fertilizer labels. For packaged products, this information shall either (1) appear on the front or back of the package, (2) occupy at least the upper-third of a side of the package, or (3) be printed on a tag and attached to the package. This information shall be in a readable and conspicuous form. For bulk products, this same information in written or printed form shall accompany delivery and be supplied to the purchaser at time of delivery.
   (a) Net Weight
   (b) Brand
   (c) Grade (Provided that the grade shall not be required when no primary nutrients are claimed.)
   (d) Guaranteed Analysis[1]
       Total Nitrogen (N)[2] .... _____%
       _____% Ammoniacal Nitrogen

---

[1]Zero (0) guarantees should not be made and shall not appear in statement.

[2]If chemical forms of N are claimed or required, the form shall be shown and the percentages of the individual forms shall add up to the total nitrogen percentage.

_____% Nitrate Nitrogen
_____% Water Insoluble Nitrogen
_____% Urea Nitrogen
_____% (Other recognized and determinable forms of N)
Available Phosphoric Acid (P_2O_5) . . . . .____%
Soluble Potash (K_2O) . . . . .____%
(Other nutrients, elemental basis)[3] . . . . .____%

  (e) Sources of nutrients, when shown on the label, shall be listed below the completed guaranteed analysis statement.

  (f) Name and address of registrant or licensee.

3. Slowly Released Plant Nutrients.

  (a) No fertilizer label shall bear a statement that connotes or implies that certain plant nutrients contained in a fertilizer are released slowly over a period of time, unless the nutrient or nutrients are identified and guaranteed.

  (b) Types of products with slow release properties recognized are (1) water insoluble (N products only), such as natural organics, ureaform materials, urea-formaldehyde products, IBDU, oxamide, etc., (2) coated slow release, such as sulfur coated urea and other encapsulated soluble fertilizers, (3) occluded slow release, where fertilizers or fertilizer materials are mixed with waxes, resins, or other inert materials and formed into particles and (4) products containing water soluble nitrogen such as ureaform materials, urea-formaldehyde products, methylenediurea (MDU), dimethylenetriurea (DMTU), dicyanodiamide (DCD), etc. The terms, "water insoluble," "coated slow release," "slow release," "controlled release," "slowly available water soluble," and "occluded slow release" are accepted as descriptive of these products, provided the manufacturer can show a testing program substantiating the claim (testing under guidance of Experiment Station personnel or a recognized reputable researcher acceptable to the _____). A laboratory procedure, acceptable to the _____ for evaluating the release characteristics of the product(s) must also be provided by the manufacturer. (Official 1987.)

  (c) To supplement (b) when the nitrogen is organic, it should be established that if a label states the amount of organic nitrogen present in a phrase, such as "nitrogen in organic form equivalent to X% N," then the water insoluble nitrogen guarantee must not be less than 60% of the nitrogen so designated. For example: If the total nitrogen guarantee for a fertilizer is 10% and the label states, "Nitrogen in organic form equivalent to 2.5% N," then the WIN guarantee must not be less than 1.5% ($2.5\% \times 0.6 = 1.5\%$).

  (d) When a slowly released nutrient is less than 15% of the guarantee for either total nitrogen (N), available phosphoric acid (P_2O_5), or soluble potash (K_2O), as appropriate, the label shall bear no reference to such designations.

  (e) Until more appropriate methods are developed, AOAC method 2.074 (13th Edition), or as it shall be designated in subsequent editions, is to be used to confirm the coated slow release and occluded slow release nutrients

---

[3]As prescribed by regulation No. 1. (Official 1986.)

and others whose slow release characteristics depend on particle size. AOAC method 2.072 (13th Edition) shall be used to determine the water insoluble nitrogen of organic materials.

4. Definitions.

Except as the _____ designates otherwise in specific cases, the names and definitions for commercial fertilizers shall be those adopted by the Association of American Plant Food Control Officials.

5. Percentages.

The term of "percentage" by symbol or word, when used on a fertilizer label shall represent only the amount of individual plant nutrients in relation to the total product by weight.

6. Investigational Allowances.

(a) A commercial fertilizer shall be deemed deficient if the analysis of any nutrient is below the guarantee by an amount exceeding the values in the following schedule, or if the overall index value of the fertilizer is below 98%[4]

Guaranteed percent	Nitrogen percent	Available Phosphoric Acid, percent	Potash percent
04 or less	0.49	0.67	0.41
05	0.51	0.67	0.43
06	0.52	0.67	0.47
07	0.54	0.68	0.53
08	0.55	0.68	0.60
09	0.57	0.68	0.65
10	0.58	0.69	0.70
12	0.61	0.69	0.79
14	0.63	0.70	0.87
16	0.67	0.70	0.94
18	0.70	0.71	1.01
20	0.73	0.72	1.08
22	0.75	0.72	1.15
24	0.78	0.73	1.21
26	0.81	0.73	1.27
28	0.83	0.74	1.33
30	0.86	0.75	1.39
32 or more	0.88	0.76	1.44

For guarantees not listed, calculate the appropriate value by interpolation.

---

[4]For these investigational allowances to be applicable, the recommended AOAC procedures for obtaining samples, preparation and analysis must be used. These are described in *Official Methods of Analysis of the Association of Official Analytical Chemists,* 13th Edition, 1980, and in succeeding issues of the *Journal of the Association of Official Analytical Chemists.* In evaluating replicate data, Table 19, page 935, *Journal of the Association of Official Analytical Chemists,* Volume 49, No. 5, October, 1966, should be followed.

The overall index value is calculated by comparing the commercial value guaranteed with the commercial value found. Unit values of the nutrients used shall be those referred to in Section 10 of the Act.

Overall index value — Example of calculation for a 10-10-10 grade found to contain 10.1% Total Nitrogen (N), 10.2% Available Phosphoric Acid (P_2O_5) and 10.1% Soluble Potash (K_2O). Nutrient unit values are assumed to be $3 per unit N, $2 per unit P_2O_5, and $1 per unit K_2O.

10.0 units N	× 3 =	30.0
10.0 units P_2O_5	× 2 =	20.0
10.0 units K_2O	× 1 =	10.0
	Commercial Value Guaranteed =	60.0

10.1 units N	× 3 =	30.3
10.2 units P_2O_5	× 2 =	20.4
10.1 units K_2O	× 1 =	10.1
	Commercial Value Found	= 60.8

Overall Index Value = 100(60.8 / 60.00) = 101.3%

(b) Secondary and minor elements shall be deemed deficient if any element is below the guarantee by an amount exceeding the values in the following schedule:

Element	Allowable Deficiency
Calcium	0.2 unit + 5% of guarantee
Magnesium	0.2 unit + 5% of guarantee
Sulfur	0.2 unit + 5% of guarantee
Boron	0.003 unit + 15% of guarantee
Cobalt	0.0001 unit + 30% of guarantee
Molybdenum	0.0001 unit + 30% of guarantee
Chlorine	0.005 unit + 10% of guarantee
Copper	0.005 unit + 10% of guarantee
Iron	0.005 unit + 10% of guarantee
Manganese	0.005 unit + 10% of guarantee
Sodium	0.005 unit + 10% of guarantee
Zinc	0.005 unit + 10% of guarantee

The maximum allowance when calculated in accordance to the above shall be 1 unit (1%).

7. Sampling.

Sampling equipment and procedures shall be those adopted by the Association of Official Analytical Chemists wherever applicable.

8. Breakdown of Plant Food Elements Within the Guaranteed Analysis.

When a plant nutrient guarantee is broken down into the component forms, the percentage for each component shall be shown before the name of the form.

EXAMPLE: 4% Nitrate Nitrogen.

UNIFORM REGISTRATION APPLICATION (8 1/2" x 11")

APPLICATION FOR REGISTRATION OF COMMERCIAL FERTILIZERS IN _____

Mail to: _____ (name and address of enforcement agency)

Application is hereby made for the registration of _____ brands and grades of commercial fertilizer at $_____ per brand and grade for the period ending _____, 19 ___.

Make check payable to _____

REGISTERED BY _____

STREET _____ CITY _____ STATE _____ ZIP CODE _____

NET WEIGHT OF PACKAGES: _____

DATE _____, 19 ___ SIGNED BY _____ TITLE _____

CERTIFICATE OF REGISTRATION

This certifies that the following brands and grades of commercial fertilizer have been approved for sale in _____, 19 ___ beginning with the date of registration and ending _____.

DATE _____, 19 ___ AMOUNT _____ BY _____

BRAND NAME AND GRADE OF FERTILIZER OR FERTILIZER MATERIAL	Nitrogen	Avail. Phos. Acid	Pot-ash	Ca	Mg	S	B	Cl	Co	Cu	Fe	Mn	Mo	Na	Zn	Registration Number

Note: See other side for minimum guarantees for secondary and micro-nutrients.

REVERSE SIDE

SECONDARY AND MICRO-NUTRIENTS: When claims for such nutrients are made, the minumum percentages which will be accepted for registration are as follows:

Calcium (Ca)................1.00
Magnesium (Mg)..............0.50
Sulfur (S)..................1.00
Boron (B)...................0.02
Chlorine (Cl)...............0.10
Cobalt (Co).................0.0005

Copper (Cu)....................0.05
Iron (Fe)......................0.10
Manganese (Mn).................0.05
Molybdenum (Mo)................0.0005
Sodium (Na)....................0.10
Zinc (Zn)......................0.05

Uniform Registration Application

APPLICATION FOR COMMERCIAL FERTILIZER LICENSE

APPLICATION FOR COMMERCIAL FERTILIZER LICENSE	
STATE OF _____ DATE: _____ 19 __	*FOR CONTROL OFFICIAL'S USE*
Control Agency	
Address	

Application is hereby made and fee of $ _____ enclosed for a license to manufacture and/or distribute commercial fertilizer at the following location for the period _____, 19 ___ through _____, 19 ___.

Applicant's Business Name	*Plant Name, If Different*
P.O. Box *Number & Street*	*P.O. Box* *Number & Street*
City *State* *Zip Code*	*City* *State* *Zip Code*
Person in Charge and Title	*Area Code Telephone Number*

Type of Operation: Manufacturer; Distributor; Other _____
Describe Kind

Materials
Produced: ☐ Dry Blends; ☐ Ammoniation; ☐ Liquid Hot Mix;

☐ Liquid Cold Mix; ☐ Anhydrous Ammonia; ☐ Solutions;

☐ Suspensions; ☐ Secdy. or Micronutrients; ☐ Other _____
Desc.

_____
Signature

Submit in duplicate. *Make check payable to:* (*Name of agency authorized to receive funds*)

COMMERCIAL FERTILIZER LICENSE

This license entitles the above named applicant to manufacture and/or distribute commercial fertilizer at the above location in the State of _____ for the period _____, 19 ___ through _____, 19 ___. Such license shall remain in effect unless suspended or revoked by the _____ for cause.

LICENSE NUMBER: _____ DATE _____, 19 ___.

_____
Signature Control Official

_____
Title

Application for Commercial Fertilizer License

APPENDIX C

Useful Tables and Conversions

Table C-1

Useful Weights and Measures for the Home Gardener and Horticulturist

Weights

Pounds per Acre		Equivalent Quantity per 100 Square Feet
100	. . . .	3½ oz.
200	. . . .	7½ oz.
300	. . . .	11 oz.
400	. . . .	14¾ oz.
500	. . . .	1 lb. 2½ oz.
600	. . . .	1 lb. 6 oz.
700	. . . .	1 lb. 10 oz.
800	. . . .	1 lb. 13 oz.
900	. . . .	2 lbs. 1 oz.
1,000	. . . .	2 lbs. 5 oz.
2,000	. . . .	4 lbs. 10 oz.

Measures
(approximate)

1 level tsp.	⅙ oz.
1 level tbsp.	½ oz.
1 level c.	8 oz.
1 pt.	1 lb.
1 qt.	2 lbs.
1 gal.	8 lbs.

Table C-2

Conversion of Acres to Smaller Units[1]

A. Conversions for flower pots or flower boxes

At These Rates Per Acre	Use These Volumes of Average Fertilizer (in tsp) Flower pots			Flower Boxes	
	4-inch	6-inch	8-inch	1 sq. ft.	2 sq. ft.
500	.05[2]	.10[2]	.30[2]	1	2
1,000	.10[2]	.20[2]	.60[2]	2	4

B. Conversions for small areas

Find Your Material in the List Below	Find the Recommended Rate per Acre	Determine the Amount of Fertilizer Needed for Your Specific Area						

Material Grouped According to Approximate Weight per Pint	Amounts per Acre	Use These Weights per Specified Area		Use These Volumes per Specified Area		10 Feet of Row Spaced[3]		
		100 sq. ft.	1,000 sq. ft.	10 sq. ft.	100 sq. ft.	1 ft.	2 ft.	3 ft.
	(Lbs.)	(Lbs.)	(Lbs.)	(Tbsps.)	(Pints)	(Tbsps.)	(Tbsps.)	(Cups)
Activated Sewage Sludge	100	.2	2.3	1.2	.4	1.2	2.4	.2
Dried Blood	500	1.2	11.5	6.0	1.9	6.0	12.0	1.1
Sulfur	1,000	2.3	23.0	12.0	3.7			
(10 oz. per pint)								
Ammonium Chloride	100	.2	2.3	.9	.3	.9	1.8	.2
Ammonium Nitrate	500	1.2	11.5	4.5	1.4	4.5	9.0	.8
Urea	1,000	2.3	23.0	9.0	2.8			
(13 oz. per pint)								
Ammonium Phosphate	100	.2	2.3	.7	.2	.7	1.4	.1
Gypsum	500	1.2	11.5	3.5	1.2	3.5	7.0	.7
Mixed Fertilizers	1,000	2.3	23.0	7.0	2.3			
Potassium Chloride								
(16 oz. per pint)								

(Continued)

Table C-2 (Continued)

Material Grouped According to Approximate Weight per Pint	Use These Weights per Specified Area			Use These Volumes per Specified Area				
						10 Feet of Row Spaced[3]		
	Amounts per Acre	100 sq. ft.	1,000 sq. ft.	10 sq. ft.	100 sq. ft.	1 ft.	2 ft.	3 ft.
	(Lbs.)	(Lbs.)	(Lbs.)	(Tbsps.)	(Pints)	(Tbsps.)	(Tbsps.)	(Cups)
Ammonium Sulfate	100	.2	2.3	.6	.2	.6	1.2	.1
Calcium Nitrate	500	1.2	11.5	3.0	1.0	3.0	6.0	.6
Mixed Fertilizers	1,000	2.3	23.0	6.0	2.0			
Superphosphate (19 oz. per pint)								
Ground Limestone	100	.2	2.3	.5	.2	.5	1.0	.1
Potassium Sulfate	500	1.2	11.5	2.5	.8	2.5	5.0	.5
(23 oz. per pint)	1,000	2.3	23.0	5.0	1.6			
	2,000	4.6	46.0	10.0	3.2			

[1]Reprinted from the U.C. Division of Ag and Natural Resources; originally printed as Leaflet No. 2285.

[2]Since it is difficult to measure this small amount, you can dissolve 1 teaspoon of fertilizer in a pint of water and add appropriate proportion to the flower pot.

[3]Since high rates are not desirable in row fertilization, they are omitted in this table.

Table C-3

Number of Trees or Plants per Unit Area

	Number per:				Number per:	
Spacing	Acre	1,000 Sq. Ft.		Spacing	Acre	1,000 Sq. Ft.
1 by 2 ft.	21,780	500		6 by 6 ft.	1,210	28
1 by 3 ft.	14,520	333		6 by 8 ft.	907	21
1 by 4 ft.	10,890	250		8 by 8 ft.	680	16
1½ by 2 ft.	14,520	333		10 by 10 ft.	436	10
1½ by 3 ft.	9,680	222		12 by 12 ft.	302	
2 by 3 ft.	7,260	167		15 by 15 ft.	194	
2 by 4 ft.	5,445	125		16 by 16 ft.	170	
3 by 4 ft.	3,630	83		18 by 18 ft.	134	
3 by 5 ft.	2,904	67		20 by 20 ft.	109	
3 by 6 ft.	2,420	56		25 by 25 ft.	70	
4 by 4 ft.	2,722	62		30 by 30 ft.	48	
4 by 6 ft.	1,815	42		40 by 40 ft.	27	

Table C-4

Conversion Factors for English and Metric Units

To Convert Column 1 into Column 2, Multiply by	Column 1	Column 2	To Convert Column 2 into Column 1, Multiply by
	Length		
0.621	kilometer, km	mile, mi.	1.609
1.094	meter, m	yard, yd.	0.914
0.394	centimeter, cm	inch, in.	2.54
	Area		
0.386	kilometer^2, km^2	mile^2, mi.^{2}	2.590
247.1	kilometer^2, km^2	acre, acre	0.00405
2.471	hectare, ha	acre, acre	0.405
	Volume		
0.00973	meter^3, m^3	acre-inch	102.8
3.532	hectoliter, hl	cubic foot, ft.^{3}	0.2832
2.838	hectoliter, hl	bushel, bu.	0.352
0.0284	liter, L	bushel, bu.	35.24
1.057	liter, L	quart (liquid), qt.	0.946
	Mass		
1.102	ton (metric)	ton (English)	0.9072
2.205	quintal, q	hundredweight, cwt (short)	0.454
2.205	kilogram, kg	pound, lb.	0.454
0.035	gram, g	ounce (avdp.), oz.	28.35
	Pressure		
14.50	bar	lb. / inch^2, psi	0.06895
0.9869	bar	atmosphere, atm	1.013
0.9678	kg (weight) / cm^2	atmosphere, atm	1.033
14.22	kg (weight) / cm^2	lb. / inch^2, psi	0.07031
14.70	atmosphere, atm	lb. / inch^2, psi	0.06805
0.1450	kilopascal	lb. / inch^2, psi	6.895
0.009869	kilopascal	atmosphere, atm	101.30
	Yield or Rate		
0.446	ton (metric) / hectare	ton (English) / acre	2.240
0.891	kg / ha	lb. / acre	1.12
0.891	quintal / hectare	hundredweight / acre	1.12
1.15	hectoliter / hectare	bu. / acre	0.87

Table C-5

The Metric System

The fundamental unit of the metric system is the meter (the unit of length), from which the units of mass (gram) and capacity (liter) are derived; all other units are the decimal subdivisions or multiples thereof. These three units are simply related, so that for all practical purposes the volume of 1 kilogram of water (1 liter) is equal to 1 cubic decimeter.

Prefix		Meaning		Units	
micro-	= one millionth,	$1/_{1,000,000.}$		0.000001	
milli-	= one thousandth,	$1/_{1,000.}$		0.001	
centi-	= one hundredth,	$1/_{100.}$		0.01	
deci-	= one tenth,	$1/_{10.}$		0.1	
unit	= one,			1.	*meter* for length
deka- or deca-	= ten,	$10/_{1.}$		10.	*gram* for mass,
hecto-	= one hundred,	$100/_{1.}$		100.	*liter* for capacity.
kilo-	= one thousand,	$1,000/_{1.}$		1,000.	
mega-	= one million,	$1,000,000/_{1.}$		1,000,000.	

The metric terms are formed by combining the words "meter," "gram" and "liter" with the eight numerical prefixes.

The finer subdivisions of measurement, nano- (10^{-9}), pico- (10^{-12}) and femto- (10^{-15}), are additional prefixes sometimes usefully employed, especially with units of mass (gram) and capacity (liter).

Table C-6

Metric – U.S. System Equivalents

Length

Metric denominations and values		Equivalents in denominations in use
myriameter =	10,000 m	= 6.2137 mi.
kilometer =	1,000 m	= 0.62137 mi. or 3,280 ft. 10 in.
hectometer =	100 m	= 328 ft. 1 in.
dekameter =	10 m	= 393.7 in.
meter =	1 m	= 39.37 in.
decimeter =	0.1 m	= 3.937 in.
centimeter =	0.01 m	= 0.3937 in.
millimeter =	0.001 m	= 0.0394 in.

Volume

Name	No. liters	Cubic measure	Dry measure	Liquid measure
kiloliter =	1,000	= 1 cu m	= 1.308 cu yds.	= 264.17 gal.
hectoliter =	100	= 0.1 cu m	= 2 bu. 3.35 pks.	= 26.417 gal.
dekaliter =	10	= 10 cu dm	= 9.08 qt.	= 2.6417 gal.
liter =	1	= 1 cu dm	= 0.908 qt.	= 1.0567 qt.
deciliter =	0.1	= 0.1 cu dm	= 6.1022 cu in.	= 0.845 gill
centiliter =	0.01	= 10 cu cm	= 0.6102 cu in.	= 0.338 fluid oz.
milliliter =	0.001	= 1 cu cm	= 0.061 cu in.	= 0.27 fluid dr.

Weight

Name	No. grams	Cubic measure[1]	Avoirdupois weight
millier or tonneau =	1,000,000	= 1 cu m	= 2204.6 lbs.
quintal =	100,000	= 1 hl	= 220.46 lbs.
myriagram =	10,000	= 10 L	= 22.046 lbs.
kilogram or kilo =	1,000	= 1 L	= 2.2046 lbs.
hectogram =	100	= 1 dl	= 3.5274 oz.
dekagram =	10	= 10 cu cm	= 0.3527 oz.
gram =	1	= 1 cu cm	= 15.432 gr.
decigram =	0.1	= 0.1 cu cm	= 1.5432 gr.
centigram =	0.01	= 10 cu mm	= 0.1543 gr.
milligram =	0.001	= 1 cu. mm	= 0.0154 gr.

Area

hectare =	10,000 sq. m	= 2.471 acres
are =	100 sq. m	= 119.6 sq. yds.
centare =	1 sq. m	= 1,550 sq. in.

[1]Based on pure water at 4°C and 760 mm pressure.

Table C-7

Temperature Comparison of Celsius to Fahrenheit

Celsius (C°)	Fahrenheit (F°)
−30	−22
−20	− 4
−10	14
0	32
10	50
20	68
30	86
40	104
50	122
60	140
70	158
80	176
90	194
100	212

Conversion Formulas

$$C° = \frac{5}{9}(F° - 32) \qquad F° = (\frac{9}{5}C°) - 32$$

Table C-8

Useful Conversions

The following data are useful in calculating rates of application:

1 acre-foot of soil = 4,000,000 lbs. (approximate)
1 t. per acre = 20.8 g per sq. ft.
1 t. per acre = 1 lb. per 21.78 sq. ft.
1 t. per acre = 25.12 quintals per hectare
1 t. per acre 6″ depth = 1 g per 1,000 g of soil
1 g per sq. ft. = 96 lbs. per acre
1 lb. per acre = 0.0104 g per sq. ft.
1 lb. per acre = 1.12 kilos per hectare
100 lbs. per acre = 0.2296 lbs. per 100 sq. ft.
g per sq. ft. × 96 = lbs. per acre
kg per 48 sq. ft. = t. per acre
lbs. per sq. ft. × 21.78 = t. per acre
lbs. per sq. ft. × 43,560 = lbs. per acre
100 sq. ft. = 1 / 435.6 or 0.002296 acre
t. per acre-foot = 0.00136 × parts per million
cu ft. per second = 0.002228 gal. per minute
parts per million = 17.1 × gr. per gal.
parts per million × 0.00136 = t. per acre-foot

Table C-9

U.S. Weights

Troy Weight

24 grains (gr.)	= 1 pennyweight (pwt. or dwt.)
20 pennyweight	= 1 ounce (oz.)
12 ounces	= 1 pound (lb.)

Apothecaries Weight

20 grains (gr.)	= 1 scruple (sc.)
3 scruples	= 1 dram (dr.)
8 drams	= 1 ounce (oz.)
12 ounces	= 1 pound (lb.)

Avoirdupois Weight

$27^{11}/_{32}$ grains (gr.)	= 1 dram (dr.)
16 drams	= 1 ounce (oz.)
16 ounces	= 1 pound (lb.)
25 pounds	= 1 quarter
4 quarters or 100 pounds (U.S.)	= 1 hundredweight (cwt.)
112 pounds (Gr. Brit.)	= 1 hundredweight
20 hundredweight or 2,000 pounds (U.S.)	= 1 ton (t.)

The U.S. short ton is 2,000 lbs., the British long ton is 2,240 lbs. and the metric ton (1,000 kg) is 2,204.6 lbs. The long ton is also used in the United States and as a measure of weight especially by steamship companies and customs officials.

Table C-10

Land Measurements

Linear Measure		Square Measure	
1 in.	0.0833 ft.	144 sq. in.	1 sq. ft.
7.92 in.	1 link	9 sq. ft.	1 sq. yd.
12 in.	1 ft.	30¼ sq. yds.	1 sq. rd.
1 vara	33 in.	16 sq. rds.	1 sq. ch.
2¾ ft.	1 vara	1 sq. rd.	272¼ sq. ft.
3 ft.	1 yd.	1 sq. ch.	4,356 sq. ft.
25 links	16½ ft.	10 sq. chs.	1 acre
25 links	1 rd.	160 sq. rds.	1 acre
100 links	1 ch.	4,840 sq. yds.	1 acre
16½ ft.	1 rd.	43,560 sq. ft.	1 acre
5½ yds.	1 rd.	640 acres	1 sq. mi.
4 rds.	100 links	1 sq. mi.	1 section
66 ft.	1 ch.	160 acres	¼ section
80 ch.	1 mi.	36 sq. mi.	1 twp.
320 rds.	1 mi.	6 mi. sq.	1 twp.
8,000 links	1 mi.	1 sq. mi.	2.59 sq. km
5,280 ft.	1 mi.		
1,760 yds.	1 mi.		
9 in.	1 span		
4 in.	1 hand		

Table C-11

Volume Measurements

Cubic Measure

1,728 cu in.	1 cu ft.
1 cu ft.	7.4805 gal.
27 cu ft.	1 cu yd.
128 cu ft. (4′ × 4′ × 8′)	1 cord (wood)
231 cu in.	1 gal.
2,150.4 cu in.	1 bu.
1.244 cu ft.	1 bu.

Liquid Measure

1 pt. (4 gills)	16 fluid oz.
1 qt. (2 pts.)	32 fluid oz.
1 gal. (4 qt.)	128 fluid oz.
1 gal. (U.S.)	0.8327 Imperial gal.
31½ gal.	1 barrel
42 gal.	1 barrel (petroleum measure)
63 gal. (2 barrels)	1 hogshead

Dry Measure

2 pts. dry	1 qt. dry
8 qt. dry	1 pk.
4 pks.	1 bu.
105 qt. dry or 7,056 cu in.	1 standard barrel

Gallons in Square Tanks

To find the number of gallons in a square or oblong tank, multiply the number of cubic feet that it contains by 7.4805.

Gallons in Circular Tanks

To find the number of gallons in a circular tank or well, square the diameter in feet, multiply by the depth in feet, and then multiply by 5.875.

Table C-12

Convenient Conversion Factors

Multiply	By	To Get
Acres	0.4048	Hectares
Acres	43,560	Square feet
Acres	160	Square rods
Acres	4,840	Square yards
Ares	1,076.4	Square feet
Bushels	4	Pecks
Bushels	64	Pints
Bushels	32	Quarts
Centimeters	0.3937	Inches
Centimeters	0.01	Meters
Cubic centimeters	0.03382	Ounces (liquid)
Cubic feet	1,728	Cubic inches
Cubic feet	0.03704	Cubic yards
Cubic feet	7.4805	Gallons
Cubic feet	29.92	Quarts (liquid)
Cubic yards	27	Cubic feet
Cubic yards	46,656	Cubic inches
Cubic yards	202	Gallons
Feet	30.48	Centimeters
Feet	12	Inches
Feet	0.3048	Meters
Feet	0.060606	Rods
Feet	⅓ or 0.33333	Yards
Feet per minute	0.01136	Miles per hour
Gallons	0.1337	Cubic feet
Gallons	4	Quarts (liquid)
Gallons of water	8.3453	Pounds of water
Grams	15.43	Grains
Grams	0.001	Kilograms
Grams	1,000	Milligrams
Grams	0.0353	Ounces
Grams per liter	1,000	Parts per million
Hectares	2.471	Acres
Inches	2.54	Centimeters
Inches	0.08333	Feet
Kilograms	1,000	Grams
Kilograms	2.205	Pounds
Kilograms per hectare	0.892	Pounds per acre
Kilometers	3,281	Feet
Kilometers	0.6214	Miles
Liters	1,000	Cubic centimeters
Liters	0.0353	Cubic feet
Liters	61.02	Cubic inches
Liters	0.2642	Gallons

(Continued)

Table C-12 (Continued)

Multiply	By	To Get
Liters	1.057	Quarts (liquid)
Meters	100	Centimeters
Meters	3.2181	Feet
Meters	39.37	Inches
Miles	5,280	Feet
Miles	63,360	Inches
Miles	320	Rods
Miles	1,760	Yards
Miles per hour	88	Feet per minute
Miles per hour	1.467	Feet per second
Miles per minute	60	Miles per hour
Ounces (dry)	0.0625	Pounds
Ounces (liquid)	0.0625	Pints (liquid)
Ounces (liquid)	0.03125	Quarts (liquid)
Parts per million	8.345	Pounds per million gallons water
Pecks	16	Pints (dry)
Pecks	8	Quarts (dry)
Pints (dry)	0.5	Quarts (dry)
Pints (liquid)	16	Ounces (liquid)
Pounds	453.5924	Grams
Pounds	16	Ounces
Pounds of water	0.1198	Gallons
Quarts (liquid)	0.9463	Liters
Quarts (liquid)	32	Ounces (liquid)
Quarts (liquid)	2	Pints (liquid)
Rods	16.5	Feet
Rods	5.5	Yards
Square feet	144	Square inches
Square feet	0.11111	Square yards
Square inches	0.00694	Square feet
Square miles	640	Acres
Square miles	27,878,400	Square feet
Square rods	0.00625	Acres
Square rods	272.25	Square feet
Square yards	0.0002066	Acres
Square yards	9	Square feet
Square yards	1,296	Square inches
Temperature (°C) + 17.98	1.8	Temperature, °F
Temperature (°F) − 32	$^5/_9$ or 0.5555	Temperature, °C
Tons	907.1849	Kilograms
Tons	2,000	Pounds
Tons, long	2,240	Pounds
Yards	3	Feet
Yards	36	Inches
Yards	0.9144	Meters

Table C-13

Essential Growth Elements, Their Atomic Weights and Common Valence Values

Name	Symbol	Atomic Weight	Common Valence
Boron	B	10.82	3
Calcium	Ca	40.08	2
Carbon	C	12.01	4
Chlorine	Cl	35.46	−1
Copper	Cu	63.54	1, 2
Hydrogen	H	1.01	1
Iron	Fe	55.85	2, 3
Magnesium	Mg	24.31	2
Manganese	Mn	54.94	2, 4, 7
Molybdenum	Mo	95.94	3, 4, 6
Nitrogen	N	14.01	3, 5
Oxygen	O	16.00	−2
Phosphorus	P	30.98	5
Potassium	K	39.10	1
Sulfur	S	32.06	4, 6
Zinc	Zn	65.37	2

Table C-14

Atomic Weights of Elements in Common Fertilizer Materials

Name	Symbol	Atomic Weight	Name	Symbol	Atomic Weight
Aluminum	Al	26.97	Magnesium	Mg	24.31
Boron	B	10.82	Manganese	Mn	54.94
Calcium	Ca	40.08	Molybdenum	Mo	95.94
Carbon	C	12.01	Nickel	Ni	58.69
Chlorine	Cl	35.46	Nitrogen	N	14.01
Cobalt	Co	58.94	Oxygen	O	16.00
Copper	Cu	63.54	Phosphorus	P	30.98
Fluorine	F	19.00	Potassium	K	39.10
Hydrogen	H	1.01	Sodium	Na	23.00
Iodine	I	126.92	Sulfur	S	32.06
Iron	Fe	55.85	Zinc	Zn	65.37

Table C-15

Chemical Symbols, Equivalent Weights and Common Names of Ions, Salts and Chemical Amendments

Chemical Symbol or Formula	Gram Equivalent Weight	Common Name
Ca^{++}	20.04	Calcium ion
Mg^{++}	12.15	Magnesium ion
Na^+	23.00	Sodium ion
K^+	39.10	Potassium ion
Cl^-	35.46	Chloride ion
NO_3^-	62.01	Nitrate ion
NH_4^+	17.03	Ammonium ion
$SO_4^=$	48.03	Sulfate ion
$CO_3^=$	30.00	Carbonate ion
HCO_3^-	61.02	Bicarbonate ion
$CaCl_2$	55.50	Calcium chloride
$CaSO_4$	68.07	Calcium sulfate
$CaSO_4 \cdot 2H_2O$	86.09	Gypsum
$CaCO_3$	50.04	Calcium carbonate
$MgCl_2$	47.62	Magnesium chloride
$MgSO_4$	60.19	Magnesium sulfate
$MgCO_3$	42.16	Magnesium carbonate
$NaCl$	58.46	Sodium chloride
Na_2SO_4	71.03	Sodium sulfate
Na_2CO_3	53.00	Sodium carbonate
$NaHCO_3$	84.02	Sodium bicarbonate
KCl	74.56	Potassium chloride
K_2SO_4	87.13	Potassium sulfate
K_2CO_3	69.10	Potassium carbonate
$KHCO_3$	100.12	Potassium bicarbonate
S	16.03	Sulfur
SO_2	32.03	Sulfur dioxide
H_2SO_4	49.04	Sulfuric acid
$Al_2(SO_4)_3 \cdot 18H_2O$	111.08	Aluminum sulfate
$FeSO_4 \cdot 7H_2O$	139.02	Iron sulfate (ferrous)

APPENDIX D

Measuring Irregularly Shaped Areas

HOW TO MEASURE AN AREA

A—Stepping Off and Calculating Approximate Areas

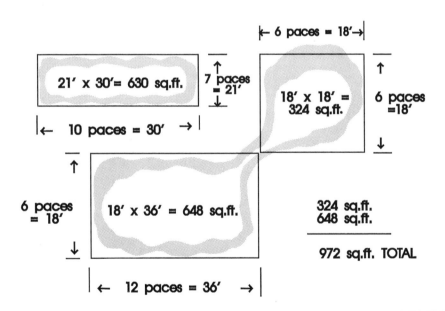

B—Comprehensive Calculations

(Portions taken from University of California Publication 4053)

Areas of turfgrass that require treatment are generally much smaller than those treated in agriculture. So, measurements, calculations and directions must be

267

followed as closely as possible when applying fertilizer in order to avoid overuse of the material. Here we explain how to calculate area measurements and how to determine fertilizer applications for different size plots when directions are given only for large acreages.

Two determinations must be made before treating any given area: One is the size of the area to be treated, and the other is the precise amount of the fertilizer to be used. Frequently unsatisfactory control is blamed on the fertilizer used when, in fact, failure is due to either wrong calculations of the size of the area to be treated or the amount of fertilizer to be applied, or both.

EXAMPLES AND CALCULATIONS

Determining the size of a given area can be simplified by dividing it into regular geometric shapes; assigning letters, such as *a, b, c,* and *d,* to represent their dimensions; and using the formulae given in this section. Generally, any area *(A)* can be considered as a square or rectangle. Odd extremities of an area can be visualized as measurable triangles or circles. For example, the fairways of a golf course can be visualized as rectangles, its tees as squares and its greens, lakes and water reservoirs as circles.

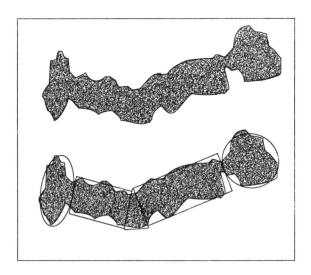

An irregular area reduced to simple geometric shapes.

SQUARE

A

20 ft. a

20 ft.

a

Formula:

$A = a \times a$, where
    a = height and width

Example: a = 20 ft.
         A = 20 ft. $\times$ 20 ft.
            = 400 sq. ft.

RECTANGLE

A

15 ft. b

40 ft.

a

Formula:

$A = a \times b$, where
    a = length, and
    b = height (or width)

Example: a = 40 ft.
         b = 15 ft.
         A = 40 ft. $\times$ 15 ft.
            = 600 sq. ft.

TRIANGLE

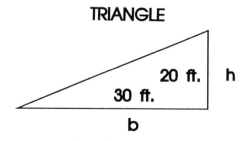

Formula:

$$A = \frac{h \times b}{2}$$

Example: $A = \dfrac{20 \text{ ft.} \times 30 \text{ ft.}}{2}$

$$= 300 \text{ sq. ft.}$$

CIRCLE

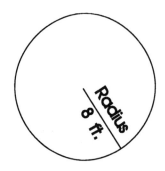

Formula:

$A = \pi r^2$, where
$\quad \pi = 3.14$, and
$\quad r = $ radius

Example: $\pi = 3.14$
$\quad r = 8$ ft.
$\quad A = 3.14 \times 8 \text{ ft.} \times 8 \text{ ft.}$
$\quad\quad = 200.96$ sq. ft.

ELLIPSE

If the geometric shape resembles an ellipse rather than a circle, the formula "$A = 0.7854 \times a \times b$" is used, with a representing the length of the ellipse and b the shorter length or what may be considered its width.

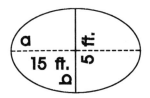

Formula:

$$A = 0.7854 \times a \times b, \text{ where}$$
$$a = \text{length of the ellipse, and}$$
$$b = \text{shorter dimension (width)}$$

Example: $a = 15$ ft.
$b = 5$ ft.
$A = 0.7854 \times 15 \times 5$
$= 58.9$ sq. ft.

IRREGULARLY SHAPED AREA

Method I. Determination of a very irregularly shaped area can be obtained by establishing the longest line possible lengthwise through the center of the area. Numerous lines are then established perpendicular to this center line. The total number of lines will depend upon how irregular the shape of the area may be. The more irregular it is, the more lines should be drawn. From the average length of all these lines, the width of the area is determined and the area calculated as a rectangle.

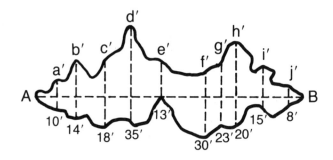

Formula:

$A = a \times b$, where

    a = distance between A and B, and

    b = average of all lengths a' to j' (lines are drawn perpendicular to a)

Example: a' = 10 ft.
           b' = 14 ft.
           c' = 18 ft.
           d' = 35 ft.
           e' = 13 ft.
           f' = 30 ft.
           g' = 23 ft.
           h' = 20 ft. a = 128 ft.
           i' = 15 ft. b = 18.6 ft. (186 ÷ 10)
           j' = 8 ft. A = 2,380.8 sq. ft. (128 × 18.6)

    Total 186 ft.

Method II. Another method for determining the size of an irregularly shaped area, a golf green, for example, is to establish a point as near to the center of the area as can be estimated. From this point, as with a compass, distances for each 10-degree increment are measured to the edge of the irregularly shaped green. Then, the 36 measurements taken completely around the central point are averaged. The idea is to obtain an average measurement, and that measurement becomes the radius of the circle. The area then is computed using the formula for a circle.

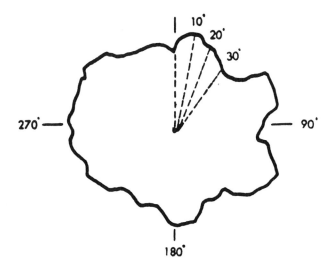

Formula:
$A = \pi r^2$

Example:

Degrees	Distance (ft): center to periphery
10 (r^1)	54.8
20 (r^2)	43.9
30 (r^3)	48.4
40 (r^4)	46.9
330 (r^{33})	41.5
340 (r^{34})	48.6
350 (r^{35})	51.0
360 (r^{36})	50.0
Total	1,980.0

$r = 1,980 \div 36 = 55$ ft.
$A = 3.14 \times 55 \text{ ft.} \times 55 \text{ ft.} = 9,498$ sq. ft.

Index

NOTES